环境工程
专业英语与写作

丁园 丁琳 主编

Environmental Engineering Professional English and Writing

化学工业出版社
·北京·

内容简介

《环境工程专业英语与写作》共分为十一个单元，涵盖了人与环境、生态学、大气污染、水污染与废水处理、固体废物及其处理、噪声及其影响、土壤污染与修复和环境监测与现代仪器分析等专题。每个单元包括课文、阅读材料、写作技巧和练习，从不同的维度锻炼读者的阅读和写作能力。本书将翻译和写作技巧等内容与课文有机衔接，有助于学生全面掌握环境工程及相关专业英语与写作的知识。

本书可作为环境科学与工程、给排水科学与工程等专业的专业英语写作教材，也可供相关专业人员参考学习。

图书在版编目（CIP）数据

环境工程专业英语与写作 / 丁园，丁琳主编. —北京：化学工业出版社，2024.6
ISBN 978-7-122-44733-3

I. ①环… II. ①丁… ②丁… III. ①环境工程-英语-论文-写作 IV. ①X5

中国国家版本馆 CIP 数据核字（2024）第 072705 号

责任编辑：汪　靓　宋林青　　　　装帧设计：史利平
责任校对：杜杏然

出版发行：化学工业出版社
（北京市东城区青年湖南街 13 号　邮政编码 100011）
印　　装：北京科印技术咨询服务有限公司数码印刷分部
710mm×1000mm　1/16　印张 9　字数 173 千字
2024 年 7 月北京第 1 版第 1 次印刷

购书咨询：010-64518888　　　　售后服务：010-64518899
网　　址：http://www.cip.com.cn
凡购买本书，如有缺损质量问题，本社销售中心负责调换。

定　　价：29.80 元

前言

环境工程专业英语是一门为普通高校环境工程及给排水科学与工程等相关专业本科生开设的语言应用课程，涉及环境工程领域庞大而复杂的专业知识。为适应互联网+时代的智慧课堂教学的需要，教材将纸质教材与在线课程网站（http://mooc1.chaoxing.com/mooc-ans/course/241627850.html）和教学资源库等线上教育资源有机衔接，以培养具有国际视野、能够用英语进行学术交流的环境工程以及给排水科学与工程专业人才。本教材着重从生态文明与可持续发展的大环保意识、爱国主义情怀和以逻辑思维和辩证思维为代表的科学素养等几个方面入手，有机融入思政元素，实现为党育人、为国育才的目标。教材在分章介绍环境工程及相关专业知识的基础上，结合了科技论文写作的知识，突出实用性和操作性，为广大学生和专业人士提供一本实用的教材。

本教材共分为十一个单元，涵盖了人与环境、生态学、大气污染、水污染与废水处理、固体废物及其处理、噪声及其影响、土壤污染与修复和环境监测与现代仪器分析等专题。每个单元包括课文、阅读材料、写作技巧、练习四个部分。课文着重介绍专业英语的典型句法与构词特点；阅读材料主要从各大数据库的环境学科类期刊中选取；本书注意将翻译技巧和写作技巧等内容与课文有机衔接，以便学生全面掌握环境工程及相关专业英语与写作的知识。附录中的必备词汇为学生提供了一份实用的词汇表，有助于学生词汇量的积累。

本教材立足于环境工程专业英语与写作的实际需求，以培养具备国际视野的环境工程专业人才为目标，在各个单元的文章选材中，我们注重反映人与自然共生共融的理念，培养学生生态文明与可持续发展的观念，我们希望本教材能为广大学生和专业人士提供有益的帮助，助力我国环境工程领域的发展。

本书由南昌航空大学丁园和丁琳主编，罗旭彪主审。在本书的编写过程中，化学工业出版社的编辑提出了宝贵意见，同时南昌航空大学在资金上给予了大力支持，在此致以诚挚的感谢。

编者

2023 年 10 月

目录

Unit 1
Environmental Science

【重点与难点】

（1）It is... that 强调句型的使用；（2）Only 强调句型的使用；（3）常用构词法（pre-, hydro-, -phere 等）；（4）双重否定的译法；（5）科技论文的修辞特点。

【学习指南】

依据课程重难点，结合课程线上资源及教材的课后习题完成课前预习和课后复习，了解环境的概念及环境工程师的使命与职业道德规范要求。

Text 1

Environment and Environmental Engineering

Simply stated, the environment can be defined as one's surroundings. In terms of the environmental engineer's involvement, however, a more specific definition is needed. To the environmental engineer, the word environment may take on global dimensions, may refer to a very localized area in which a specific problem must be addressed, or may, in the case of contained environments, refer to a small volume of liquid, gaseous, or solid materials within a treatment plant reactor. ①

The global environment consists of the atmosphere, the hydrosphere, and the lithosphere in which the life-sustaining resources of the earth are contained. The atmosphere, a mixture of gases extending outward from the surface of the earth, evolved from elements of the earth that were gasified during its formation and metamorphosis. ② The hydrosphere consists of the oceans, the lakes and streams and the shallow ground water bodies that interflow with the surface water. The lithosphere is the soil mantle that wraps the core of the earth.

The biosphere, a thin shell that encapsulates the earth, is made up of the atmosphere and lithosphere adjacent to the surface of the earth, together with the hydrosphere. It is within the biosphere that the life forms of earth, including humans, live. Life-sustaining materials in gaseous, liquid, and solid forms are cycled through the biosphere, providing substance to all living organisms.

Life-sustaining resources, air, food, and water, are withdrawn from the biosphere. It is also into the biosphere that waste products in gaseous, liquid, and solid forms are discharged.[③] From the beginning of time, the biosphere has received and assimilated the wastes generated by plant and animal life. Natural systems have been ever active, dispersing smoke from forest fires, diluting animal wastes washed into streams and rivers, and converting debris of past generations of plant and animal life into soil rich enough to support future populations.

For every natural act of pollution, for every undesirable alteration in the physical, chemical, or biological characteristics of the environment, for every incident that eroded the quality of the immediate, or local environment, there were natural actions that restored that quality.[④] Only in recent years has it become apparent that the sustaining and assimilative capacity of the biosphere, though tremendous, is not, after all, infinite. Though the system has operated for millions of years, it has begun to show signs of stress, primarily because of the impact of humans upon the environment.

Environmental engineering has been defined as the branch of engineering that is concerned with protecting the environment from the potential, deleterious effects of human activity, protecting human populations from the effects of adverse environmental factors, and improving environmental quality for human health and well-being.

As the above definition implies, humans interact with their environment—sometimes adversely impacting the environment and sometimes being adversely impacted by pollutants in the environment. An understanding of the nature of the environment and of human interaction with it is a necessary prerequisite to understanding the work of the environmental engineer.[⑤]

New Words and Expressions

environmental	*adj*. 环境的，环境产生的
engineering	*n*. 工程，工程学
simply stated	简单地说
define	*vt*. 解释，给…… 下定义，明确说明，限定
be defined as	定义为……

in terms of	在……所指的范畴，用……术语
take on	采取，呈现，承受
refer to	指……
engineer	*n.* 工程师，轮机员；*vt.* 督造，指导
in the case of	在……情况下
contained	*adj.* 封闭的，包含的
involvement	*n.* 包含，卷入，连累，牵连，困境
dimension	*n.* 面积，容积，大小，尺寸，维
localize	*vt.* 局部化，地方化
treatment-plant	处理厂
reactor	*n.* 电抗器；反应堆；反应器
evolve	*vt.* 向进，发展，演变，推论，引申，（使）发展
evolved from	从……演变而来
hydrosphere	*n.* 水圈
lithosphere	*n.* 岩石圈
life-sustaining	维持生命，支撑，维持，供养，使生存下去
metamorphosis	*n.*（复数-es）变化（形、性、质、态）
groundwater	*n.* 地下水
interflow	*vi.*；*n.* 补给，流通
interflow with	与……互相流通
mantle	*n.* 披风，覆盖物；*vt.* 覆盖，罩上
core	*n.* 核心
biosphere	*n.* 生物圈
encapsulate	*vt.* 密封；用胶囊装
be made up of	由……组成
adjacent	*adj.* 接近的，临近的
gaseous	*adj.* 气的，似气体的
organism	*n.* 生物，有机体
withdraw	*vt.* 回收，缩回，撤回，退回，取消，抽出，提取
(be) withdraw(n) from	从……取来
waste	*adj.* 无用的；*vt.* 浪费，消耗，荒芜；*n.* 浪费，废物
discharge	*vt.* 放出；*n.* 流出，放出，排放，排泄物
(be) discharge(d) into	排入……
assimilate	*vt.* 同化，吸收，消化
disperse	*vt.* 驱散，传播，消散，扩散

dilute	*vt.* 稀释，冲淡
debris	*n.* 残骸
alteration	*n.* 改变，变更
biological	*adj.* 生物学的
incident	*n.* 事件，小事，事变
erode	*vt.* 侵蚀，腐蚀
immediate	*adj.* 立刻的，直接的
apparent	*adj.* 显然的，表面上的，外表的，明显的
stress	*n.* 压力，重点，强调；*vt.* 着重，强调，应力，受力状态
impact	*n.* 碰撞，冲击，效果，影响
the impact of M upon N	M 对 N 的影响
protect	*vt.* 保护
protect …from …	保护……免于……
potential	*adj.* 可能的，潜在的，势的，位的
deleterious	*adj.* 有害的，有害杂质的
adverse	*adj.* 不利的，反对的
well-being	*n.* 幸福，福利，健康
imply	*vt.* 意指，含……的意思，应用，提交
interact	*vt.* 相互作用，互相反应
prerequisite	*n.* 先决条件，前提；*adj.* 先决的，必备的
(be) concern(ed) with	与……有关，对……关切

Notes

① may take on global dimensions, may refer to a very localized area, …, or may, … refer to a small volume of … a treatment plant reactor…

可能指全球范围，也可能指非常特定的区域，或者……也可能指反应器内的一点液体、气体或固体。

② The atmosphere, a mixture of gases extending outward from the surface of the earth, evolved from elements of the earth that were gasified during its formation and metamorphosis.

大气圈是地球表面向外延伸的混合气体，由一些在形成和变化过程中气化的地球元素演化而成。

③ It is also into the biosphere that waste products in gaseous, liquid, and solid forms are discharged.

所有的气态、液态和固态的废弃物也排入了**生物圈**。

④For every natural act of pollution, for every undesirable alteration in the physical, chemical, or biological characteristics of the environment, for every incident that eroded the quality of the immediate, or local environment, there were natural actions that restored that quality.

对每一件自然污染行为，对每一件环境中令人讨厌的物理、化学和生物特性的变化，对每一件破坏直接环境或局部环境的事件，（环境）本身都有自净（能力）。

⑤ An understanding of the nature ... engineer.

其中 it 系指 environment，即了解环境的本质以及人类与环境的相互作用，是了解……的先决条件。

Reading material 1

Introduction to Environmental Science

Environmental science is the discipline that is concerned with identifying and diagnosing environment impacts. Environmental scientists first try to understand the patterns or impact or change in the natural environment caused by various human activities. Once, they understand what is occurring, environmental scientists then search for the specific cause or causes. Often, they can also get involved in seeking solutions as well.

While environmental science is critical to understanding the impact of human activities on the natural environment, societies often turn to environmental policy, environmental education, and environmental technology for implementing solutions. Both environmental policy and education are concerned with changing human behavior. Environmental policy does so in a more direct or controlling manner. The Clean Air Act, for example, specifies the allowable levels of certain kinds of gases, which can be in industrial facilities. Environmental education, on the other hand, seeks to change human behavior in more subtle ways. Educating the average consumer about the effects of air pollution from automobiles, for example, may lead some individuals to change their behavior and use less polluting forms of transportation such as walking, bicycle, or public transportation.

Lastly, environmental technology refers to solving environmental problems by using or substituting tools, techniques, or processes that have less environmental impact. For example, probably the most well known type of environmental technology is the catalytic converter, which is attached to the exhaust system and neutralizes the gases that are emitted by the engine when gasoline is burned or combusted. To solve a specific environmental problem, societies often turn to environmental policy, education or technology, or a combination of any or all of the three.

To better understand the natural environment, the impacts that humans are having on the environment, and ways in which humans can alter their behavior and technologies to reduce environmental impact, it is useful to think of nature or natural environment from a "systems" perspective. A system can be viewed as a group of interacting, interrelated, or interdependent elements forming or regarded as forming a collective entity.

Think of the natural environment as a system, which is composed of four parts or components, each with its own unique form, arrangement, characteristics and dynamic. These four subsystems include:

Atmosphere—blanket of gases that surrounds the earth or the gaseous earth.

Lithosphere—the solid earth, composed of rocks and minerals.

Hydrosphere—waters of the earth or liquid of the earth.

Biosphere—living earth, composed of plants, animals, insects, and all living things except humans.

Remember we said that our language suggests that human are not part of the natural environment? Humans make up their subsystem, know as the sociosphere, which includes all people on the earth and all human activity. Kenneth Boulding, a well-known economist, described the sociosphere as "The social system consists of all human beings on the planet and all their interrelationships, such as kinship, friendship, hostility, status, exchange, money flows, conversations, information, outputs and inputs, and so on. It includes likewise the contents of every person's mind and the physical surroundings, both natural and artificial, to which he relates. This social system clings to the surface of the earth, so that it may appropriately be called the biosphere, even though small fragments of it are now going out into space. The sociosphere thus takes its place with the lithosphere, the hydrosphere, the atmosphere, the biosphere and so on, as one of the system which enwrap this little globe. It has strong inter-relationships with the other spheres with which it is mingled and without which it could not survive. Nevertheless, it has a dynamic and an integrity of its own. It is rather thin in Antarctica, although present there; it is very dense in New York. It is a network rather than a solid sphere or shelf, yet no part of the earth's surface is very far from it. It is a system of enormous complexity, yet not wholly beyond our comprehension."

Humans have impacted the environment for a long time. Some of this impact is deliberate. Clearing a grass land to plant crops is a deliberate alteration of the environment and if the decision is whether to have food to eat or starve, or alter the natural environment, this is a pretty easy decision for most, if not all, humans to make. However, at the same time, there may be unintended environmental impacts on clearing fields. Depending on the slope of the ground, valuable topsoil may run off into nearby waterways, thereby over time making the field less productive for food and possibly choking waterways with sediment.

To respond to the various impacts that environmental change and pollution have had on the natural environment at local, urban, regional and global scales, a new mode of human existence has been suggested. This new mode seeks to provide for the needs of the current generation of humans without compromising the ability of future generations to meet their own needs and is known as sustainable development. As described in the 1987 publication “our Common Future”, sustainable development is a process of change in which policy and institutional adjustments, technological development, and the direction of investments are harmonized with the exploitation of resources.

写作技巧

科技文献的修辞特点

科技论文一般有 4 种话语结构：说明（exposition）、叙事（narration）、描述（description）和议论（argumentation）。通常在学术刊物上发表的科技论文按其性质可分为科技专论型、科技综述型、实验研究型和科普型四类。科技文体不同于文学文体，从词汇到句子结构都有许多不同之处。美国伊利诺伊大学和伊朗德黑兰大学的科技英语学者在德黑兰大学的生物、物理等十个系进行过一次科技英语词汇复现率的调查。他们通过有关科技文献 11 万词次的调查，选出 4178 个词，并划分为纯科技词、次科技词和功能词三类。统计表明：复现率最高的词是从中选出的 400 个功能词，即语言共核的基本词汇，如：high, low, make, increase, decrease 等。次科技词则是用语言共核组成的基本专业术语，如：frost-free 无霜期，sewage sludge 污水污泥，wastewater 污水等。科技词，如 photosynthesis 光合作用，chromatography 色谱。科技文体在句法上和修辞上的特点，也是在语言共核的基础上形成的。因此，我们应当在研究语言共核的基础上来研究各种文体的差异，这样往往能起到事半功倍的效果。

科技文体是一种较正式、较庄重的书面文体，与日用文体在词汇、语法和修辞上具有明显的差异，概括起来有以下特点：

1. 用词专一，多用术语(specific scientific names)

科技文体常用专业术语。这是因为专业术语一词一义，词义专一，能够确切、准确地表达科学含义。尽管专业术语往往词形较长，发音较难，语音缺乏节奏，但却广泛应用于科技文献，以代替那些词形悦目、语音动听、一词多义、生动感人的普通词汇。例如：

日用英语：Thus a great change happens to the frog during its developments; i.e. it is an animal which begins life in water but finishes it on land.

科技英语：Thus the frog undergoes a metamorphosis during its development; i.e.it is an amphibian.

2. 用词庄重，源出拉丁(formal words of Latin origin)

科技文体常以正式词语代替非正式词语，以“大”词代替“小”词，以源出拉丁语和希腊语的词代替普通词，使得文体庄重、正式。拉丁语和希腊语是现代科技英语词汇的基础。据美国科技英语学者 Oscar E. Nybaken 调查，在 1 万个最普通的词汇中，约有 46%源出拉丁语，7.2%源出希腊语，而这个比例在科技词汇中更高。专业性越强，这个比例就越高。由于拉丁语和希腊语词汇不会发生词形和词义的变化，准确稳定，所以成了创造新科技词汇的重要源泉，在科技英语中具有重要意义。例如：

日用英语：The air surrounding the earth evolved from the elements that were gasified during its change.

科技英语：The atmosphere evolved from elements of the earth that were gasified during its formation and metamorphosis.（见 Unit 1）

3. 用词精确，避免歧义(precise words without different meanings)

科技文体用词准确、简练，不用模糊词语，以避免产生歧义。日用文体所用词语虽然也能表达科学内容，但从学术角度来看往往不够精确，而且在表达方式上也比较累赘，读起来不够正式。例如：

日用英语：If the parent plant is destroyed, the young ones have a better chance of staying alive.

科技英语：If the parent plant is destroyed, the offspring have a better chance of survival.

4. 科技词汇，多方构成(scientific words of different structures)

科技词汇面广量大，构成手段各式各样：

- 普通词汇转入科技词汇

Normal 的普通词义是“正常”，当用作数学术语时常表示“法线”的含义，当用作化学术语时有“当量”的含义。

- 缩略词

  科技文献中大量使用缩略词，常用的缩合法有：

  开首字母缩合法 r.p.m = revolution per minute

  开首字母连写法 radar = radio detecting and ranging

  半缩略词 bomb = atom-bomb, U-boat= undersea-boat

  切割缩略词 mol. = molecular, wt. = weight.

  外来缩略词：来源于拉丁语的缩略词在科技文献中频繁出现，例如：ibid. = ibidem，出处同上，同前，在同一书；i.e. = that is，那就是；e.g. =exempli gratia；vs. = versus。

- 词缀构词

  借助词缀构成科技词，例如：biosociology, biosynthesis, sociosphere。

- 元音字母连成词

  借助元音字母 o 可使两部分连成一个新科技词，例如：electromagnet, medicoathletic。

- 借助旧词造新词

  例如：telecast, telebanking。

- 组合词

  几个词组合在一起构成一个科技词，例如：hand-operated, rice thresher。

5. 多用短语，结构简洁(more phrases instead of clauses)

科技文体常用各种短语，如介词短语、形容词短语、分词短语、动名词短语、动词不定式短语、独立主格结构等，尤其常用形容词短语、分词短语和介词短语充当后置定语，以代替日用文体中的从句或并列句。这样可以使句子结构简洁、明快，信息容纳量深广、丰厚。例如：

The increasing scale and scope of the experiments needed to test new hypotheses and develop new techniques and industrial processes, they have to set up research groups or teams using high-complicated equipment in elaborately-designed laboratories.

The biosphere, a thin shell that encapsulates the earth, is made up of the atmosphere and lithosphere adjacent to the surface of the earth, together with the hydrosphere.（见 Unit 1）

6. 被动语态，客观论证(objective statements in passive voice)

被动语态句在科技英语中的使用比在日用英语中更为广泛，尤其是省略施动者的被动语态句更是常见，可以说它是英语科技文体的特征句型。根据语言学家统计，在科技英语文献中，约有 1/3 的动词使用被动语态，这是因为科技文献的作者和读者所要介绍和了解的是科学事实本身，而不是做这件事情的人，即施动者。被动语态能够准确、客观、不带感情色彩地描述事实和过程，这正是科技文献所需要的。

Simply stated, the environment can be defined as one's surroundings.（见 Unit 1）

The global environment consists of the atmosphere, the hydrosphere, and the lithosphere in which the life -sustaining resources of the earth are contained.（见 Unit 1）

7. 修辞简朴，文风端正（formal in style and plain in rhetoric）

词汇朴素是科技文献写作的重要特点。科技文献一般不使用过分生僻和夸张的词，不使用那种转弯抹角使人难以找到中心的句子结构。

日用英语：The decision to eschew an immediate price increase has been taken by the firm in the interests of facilitating agreement within the industry as a whole on a standardized policy of simultaneous action on prices, which will, it is hoped, be agreed on within the next few months.

科技英语：The firm has decided to postpone an immediate price increase by a few months. It is hoped that within this period, the industry as a whole will arrive at an agreement about simultaneous action on prices.

8. 头重脚轻，割裂平衡(beginning-weight structure)

科技文献，往往信息集中在主语，谓语很简单，句子变得“头重脚轻”，这种句型在科技论文中尤为常见。有时有的作者为了避免这种现象，使句子平衡，采用成分割裂方法。例如：

Several descriptive papers which give adequate background and experimental results have been published.

In terms of the environmental engineer's involvement, however, a more specific definition is needed.（见 Unit 1）

Exercises

Ⅰ. Fill in the blanks or answer true or false.

1. In environmental engineering's specific definition, the environment can be defined as one's surroundings.
2. Some surface water may come from ground water and v. v. (vice-versa).
3. Man and his environment are interacted with each other.
4. The biosphere consists of atmosphere and lithosphere.
5. Natural environment has a capability of self purification.
6. The environmental capability is not infinite.
7. The environmental engineers' is to protect____, protect ____ and improve ____.

8. The life sustaining resources of the earth are contained in ____.

9. The elements in atmosphere came from ____.

10. Life-sustaining resources are ____ from the biosphere, into which waste products are also ____.

Ⅱ. Answer the following questions according to the text.

1. What is the definition of sustainable development?

2. What is the topic of each paragraph of this text?

3. Point out typical sentences in the text: "It is…that…" and "Only…".

Ⅲ. Translate the following sentences.

1. It was the laboratory that we built in less than five months last year.

2. Only in this way, can we have a good command of English.

3. It is stability that destroys people's ambition and barricades people's steps.

4. Only through advanced algorithms can we achieve accurate weather predictions.

5. Only when resources are exhausted will the importance of conservation be realized.

6. 正是地心引力使卫星绕着地球运转。

7. 只有通过工业界、科学家和政府部门的共同努力，问题才能最终得到解决。

8. 正是越来越多化石燃料的燃烧，导致二氧化碳排放量的增加。

9. 值得注意的是，这个系统运行状态良好。

10. 正是新型纳米材料的发展使太阳能收集领域发生了革命性的变化。

Ⅳ. Rewrite the following simple sentences into a compound sentence.

1. The global environment consists of the atmosphere, the hydrosphere, and the lithosphere.

The life-sustaining resources of the earth are contained in the global environment.

2. The atmosphere is a mixture of gases extending outward from the surface of the earth.

The atmosphere evolved from elements of the earth.

The elements of the earth were gasified during its formation and metamorphosis.

3. The biosphere is made up of the atmosphere and lithosphere adjacent to the surface of the earth, together with the hydrosphere.

The biosphere is a thin shell.

The thin shell encapsulates the earth.

4. Life-sustaining materials are cycled through the biosphere.

Life-sustaining materials are in gaseous, liquid, and solid forms.

The biosphere provides sustenance to all living organisms.

Ⅴ. Scan the QR code to read the article "Lanzhou Oxygen Study is a Breath of Fresh Air", and then discuss the mission and professional ethics of environmental engineers.

Unit 2

Environmental Problems

【重点与难点】

（1）表示否定的前缀；（2）so…that；too…to 等典型句型的用法；（3）被动句的译法。

【学习指南】

依据课程重难点，结合课程线上资源及教材的课后习题完成课前预习和课后复习，了解环境问题的起源、分类和危害，培养环境工程师的创新意识、质量意识和环保意识。

Text 2

Impact of Humans upon Environment

In a natural state, earth's life forms live in equilibrium with their environment. The numbers and activities of each species **are governed by** the resources available to them. Species interaction is common, with the waste product of one species often forming the food supply of another. Humans alone have the ability to gather resources from beyond their immediate surroundings and process those resources into different, more versatile forms. These abilities have made it possible for human population to thrive and flourish beyond natural constraints. But the natural and manufactured wastes generated and released into the biosphere by these increased numbers of human beings have upset the natural equilibrium.

Anthropogenic, or human-induced, pollutants have overloaded the system. The overloading came relatively late in the course of human interaction with the environment, perhaps because early societies **were** primarily **concerned** with meeting natural needs, needs humans share in common with most of the higher mammals. ① These peoples had not yet begun to be concerned with meeting the acquired needs associated with more advanced civilizations. ②

Early humans used natural resources to satisfy their needs for air, water, food, and shelter. These natural, unprocessed resources were readily available in the biosphere, and the residues generated by the use of such resources were generally compatible with, or readily assimilated by the environment. Primitive humans ate plant and animal foods without even disturbing the atmosphere with the smoke from a campfire. Even when use of fire became common, the relatively small amounts of smoke generated **were** easily and rapidly **dispersed** and **assimilated by** the atmosphere.

Early civilizations often drank from the same rivers in which they bathed and deposited their wastes, yet the impact of such use was relatively slight, as natural cleansing mechanisms easily restored water quality. These early humans used caves and other natural shelters or else fashioned their homes from wood, dirt, or animal skins. Often nomadic, early populations left behind few items that **were** not readily **broken down** and **absorbed by** the atmosphere, hydrosphere, or lithosphere. ③ And those items that **were** not **broken down** with time were so few in number and so innocuous as to present no significant solid waste problems.

Only as early peoples began to gather together in larger, more or less stable groupings did their impact upon their local environments begin to be significant. In 61 A. D., cooking and heating fires caused air pollution problems so severe that the Roman philosopher Seneca complained of "the stink of the smoky chimneys". By the late eighteenth century, the waters of the Rhine and the Thames had become too polluted to support game fish. From the Middle Ages the areas where food and human waste **were dumped** harbored rats, flies, and other pests.

But these early evidences of pollution overload were merely the prelude to greater overloads to come. With the dawn of the industrial revolution, humans were better able than ever to satisfy their age-old needs of air, water, food, and shelter. Increasingly, they turned their attention to other needs beyond those associated with survival. By the late nineteenth centuries, automobiles, appliances, and processed foods and beverages had become so popular as to seem necessities, and meeting these acquired needs had become a major thrust of modern industrial society.

Unlike the natural needs discussed earlier, acquired needs **are** usually **met** by items that must **be processed or manufactured or refined**, and the production, distribution, and use of such items usually results in more complex residuals, many of which are not compatible with or readily assimilated by the environment.

Take for example, a familiar modern appliance—the toaster. The shell and the heating elements are likely to be made of steel, the handle of the lift lever of plastic. Copper wires and synthetic insulation **may be used** in the connecting cord, and rubber **may be used** on the plug. In assessing the pollutants generated by the manufacture and sale of this simple appliance, it would be necessary to include all the resources expended in the mining of the metals, the extracting and refining of the petroleum, the shipping of the various materials, then the manufacturing, shipping, and selling of the finished product. The potential impact of all of these activities upon air and water quality is significant. Furthermore ,if the pollution potential involving the manufacture and use of the heavy equipment needed for the extraction and processing of the raw materials used in the various toaster components is considered, the list could go on ad nauseam. And the solid-waste disposal problems that arise when it is time to get rid of the toaster become a further factor.

As a rule, meeting the acquired needs of modern societies generates more residuals than meeting natural needs, and these residuals are likely to be less compatible with the environment and less likely to **be** readily **assimilated** into the biosphere. As societies ascend the socioeconomic ladder, the list of acquired needs, or luxuries, increases, as do the complexity of the production chain and the mass and complexity of the pollutants generated.[④] Consequently, the impact of modern human populations upon the environmental is of major concern to the environmental engineer.

New Words and Expressions

equilibrium　*n.* 平衡，均衡

in equilibrium with …　与……平衡

species　*n.* 种

govern　*vt.* 统治，管理，控制，支配

versatile　*adj.* 兴趣广泛的，多才多艺的，多功能的

it is possible for M to …　对 M 说来……是可能的

thrive　*vi.* (throve, thriven) 兴盛，成功

flourish　*vi.*; *vt.* 茂盛，兴隆，挥舞；*n.* 夸张动作，修饰

manufacture　*vt.*（大量）制造，加工

release　*vt.* 解开，解放，发布，发行；*n.* 释放，排放，表达

release into …　释放至……

anthropogenic　*adj.* 人类产生的

human-induced　*adj.* 人引起的，人招致的

overload	*vt.* 使超载，使装载过重
acquire	*vt.* 求得，获得，学得
associate	*vt.* 联合，联想
associate with …	联同（合）……，伴随
civilization	*n.* 文明，文化，文明社会（国家）
shelter	*n.* 遮蔽物，遮蔽，掩护，住房
unprocessed	*adj.* 未加工的
residue	*n.* 剩余物，残余
be compatible with…	与……是一致的（相容的）
disturb …with…	用……扰动（干扰）
campfire	*n.* 营火
fashion	*n.* 样子，方式，风尚，时尚，风气；*vt.* 制作，加工
nomadic	*adj.* 游牧的
item	*n.* 条，条款，一条（新闻），一项，一件
absorb	*vt.* 吸收，吸引……注意力
innocuous	*adj.* 无害的
Roman	*adj.* 罗马的，罗马人的，罗马天主教的
philosopher	*n.* 哲学家
stink	*vt*; vi. 发恶臭，坏透；*n.* 恶臭，难闻的气味
chimney	*n.* 烟囱，灯罩
game	*n.* 游戏，比赛，(用复数) 竞赛会，运动会；*adj.* 勇敢的
Middle Ages	*adj.* 中年的，中世纪
dump	*n.* 丛，树丛，(一) 大块
harbor	*n.* 海港；*vt.* 庇护，怀有
pest	*n.* 害物，害虫，瘟疫
evidence	*n.* 证据，证物，形迹，迹象
prelude	*n.* 序曲，序幕；*vt.* 前奏，成为……序页
dawn	*n.* 黎明，开端；*v.* 破晓，显露，出现，渐被理解
appliance	*n.* 用具，器械
beverage	*n.* 饮料
necessity	*n.* 必要，需要，必需品
thrust	*vt.* 推，冲，插入，刺，戮；*n.* 刺，要点
refined	*n.* 提炼后的
toaster	*n.* 烤面包片机
lever	*n.* 杠杆

synthetic *adj*. 合成的；*n*. 合成物
insulation *n*. 隔离，绝缘
plug *n*. 塞子，插头；*vt*. 塞，苦干，刻苦
assess *vt*. 评价，估计，征收
expend *vt*. 花费，耗费
petroleum *n*. 石油
ad nauseam 达到令人讨厌（厌烦，作呕）的程度
get rid of 除去，扔掉
ascend *vt*. 登上，上升，攀登，追溯
socioeconomic *adj*. 社会经济的
ladder *n*. 阶梯
luxury *n*. 奢侈（品），丰富
complexity *n*. 错综复杂，合成，组成
upset *v*. 破坏，打翻，扰乱
mammals *n*. 哺乳动物
readily *adv*. 轻而易举地
compatible *adj*. 相容的，一致的
primitive *adj*. 原始的
dirt *n*. 污物，废屑，淤泥，碎石，杂物；*vt*. 弄污
deposit *v*. 存，贮存
survival *n*. 生存，幸存（者）
handle *n*. 把手，柄，拉手
likely to be … 可能……，大概会……

Notes

① … natural needs, needs humans share in common with most of the higher mammals.

needs——必需品，后一个 needs 是进一步补充和说明前一个“needs”的。意为，人类和大多数较高的哺乳动物共同分享的那些必需品。

② These people had not yet… civilizations.

其中 acquired needs 是指后来学到的或后来随着更高级的文明和进步增长起来的必需品，即习得需求。

③ Often nomadic, early populations left behind few items that…

其中 early 是形容词,与 nomadic 同样是形容 populations 的，意为早期的游牧民

族往往遗弃下一些东西。

④ As societies ascend... of the pollutants generated.

其中“do”代替前面出现的动词“increases”。

ascend the socioeconomic ladder，攀登上社会经济发展的梯子，意即社会经济的高速发展。

全句意思：随着社会经济的发展，日益增多的必需品或奢侈品（的名单）在增加，而生产环节和物质的复杂性以及所生成的污染物的复杂性也在增加。

Reading Material 2

Addressing Context Dependence in Ecology

Context dependence is widely invoked to explain disparate results in ecology. It arises when the magnitude or sign of a relationship varies due to the conditions under which it is observed. Such variation, especially when unexplained, can lead to spurious or seemingly contradictory conclusions, which can limit understanding and our ability to transfer findings across studies, space, and time. Using examples from biological invasions, we identify two types of context dependence resulting from four sources: mechanistic context dependence arises from interaction effects, and apparent context dependence can arise from the presence of confounding factors, problems of statistical inference, and methodological differences among studies. Addressing context dependence is a critical challenge in ecology, essential for increased understanding and prediction.

The prevalence and problem of context dependence

Ecological studies examining the same question or process often reach different conclusions. In invasion ecology, for example, studies have found that the phylogenetic relatedness of alien to native species can inhibit or facilitate invasion, the relationship between native and alien species richness can vary from positive to negative, and the effect of disturbance on invasion is equivocal and inconsistent. When studies addressing the same question reach different conclusions, the different outcomes are often attributed to context dependence. Context dependence, or contingency, refers to situations where relationships vary depending on the conditions—the context—under which they are observe. It includes situations where the magnitude (strength) or sign (direction) of a relationship differs under different biotic, abiotic, spatiotemporal, or observational circumstance. Context dependence is commonly and increasingly invoked in ecology, as well as in other fields typified by high complexity, large scales, and heavy reliance on observational studies, such as conservation biology, evolutionary biology, and epidemiology. Context dependence provides a convenient shorthand to describe variation within and between (potentially myriad) studies and is reported across all study type. However, because context dependence can result from many processes (as discussed below), unless the underlying causes are identified, concluding that outcomes are context dependent provides little insight by itself. Furthermore, because "context dependence" is often used to describe disparate findings, widespread use of the term could suggest that there are few general principles in ecology, that ecological relationships are largely unpredictable, and that ecological phenomena can only be understood on a case-by-case basis. Here we argue that researchers can gain greater insight into ecological processes if they recognise the different sources of context dependence and account for them in the design, interpretation, and communication of their studies. We define context dependence and propose a novel typology based on two types and four sources of context dependence. We illustrate our typology using examples from biological invasions, a field where context dependence is prominent and widely discussed, but we propose that the typology is applicable across all areas of ecology. We outline steps for addressing the different types and sources of context dependence. By understanding ways in which context dependence can arise, ecologists should be better placed to distinguish fundamental (mechanistic) from apparent context dependence, increasing predictive understanding in ecology.

写作技巧

被动句的译法

叙述句的主语是施动者时，称为主动句；而主语是受动者的句子则称为被动句。汉语中很少使用被动句，有人统计，《水浒传》全书仅用了120个被动句，而在英语科技文献中，被动语态形式则较多。本节将主要介绍被动句的几种常用的翻译方法，仅供同学参考。

1. 译成主动句式

（1）原主语仍为主语

当被动句的主语为无生命的名词，又不出现by引导的行为主体，可译成汉语的主动句，原句主语仍为主语，实际上是省略“被”的汉语被动句。该法最符合汉语习惯，实际应用频率最高。

Simply stated, the environment can be defined as one 's surroundings.（见Unit 1）

译文：简言之，环境可定义为实体周围的事物。

Solution to the problem was ultimately found.

译文：这个问题的解决办法终于找到了。

（2）变状语为主语，把原主语转换为宾语

当英语被动句中出现 by 引导的行为主体或有逻辑意义上的行为主体时（通常为介词短语），可用此法。此法相当于是对标准英语被动句的转译，基本上是原句倒转，使之更符合汉语表达习惯，使用频度较高。

Friction can be reduced and the life of the machine prolonged by lubrication.

译文：润滑能减少摩擦，延长机器寿命。

The numbers and activities of each species are governed by the resources available to them.（见 Unit 2）

译文：每个物种可获得的资源影响它们的数量和活动。

（3）增益适当的主语使译文通顺

如原句无表示施动者的状语,考虑到汉语的表达习惯,可适当增加逻辑主语。

Magnesium is found the best material for making skyrocket.

译文：人们发现，镁是制造焰火的最好材料。

The world supply of low-sulphur coal is believed to be limited.

译文：人们认为，全世界低硫煤的供给有限。

由 it 作形式主语的被动句型，有时可加不确定主语。如：It is generally accepted that .../大家公认……；it is suggested that ... /有人建议……。

（4）译成汉语的无主句

原句本不愿或无法说出动作的发出者时，可译成汉语的无主语句，而把原句主语译成宾语。

Many strange new means of transport have been developed in our century, the strangest of them being perhaps the hovercraft.

译文：在本世纪，发明了许多新奇的交通工具，其中，最新奇的也许数气垫船。

Among other things, the contents of the metals in food, human tissues and body fluids must be monitored.

译文：此外还必须监测在食物、人体组织和体液中金属的含量。

2. 译成汉语判断句

英语中有些被动句中具形容性的过去分词起一种描述事物的过程、性质和状态的作用，实际上与系表结构很相近，因此转译成汉语的“是……的”的结构。

The biosphere is made up of the atmosphere and lithosphere adjacent to the surface of the earth, together with the hydrosphere.（见 Unit 1）

译文：生物圈是由大气圈、近地表的岩石圈与水圈组成的。

The overloading came relatively late in the course of human interaction with the environment, perhaps because early societies were primarily concerned with meeting natural needs, needs humans share in common with most of the higher mammals.（见 Unit 2）

译文：超载滞后于人类与环境的相互作用过程，可能是因为早期社会主要是满足必需品的需求，即人类与其他大多数较高等的哺乳动物共同分享的那些必需品。

3. 译成汉语的被动句

当某些被动表达在现代汉语中已成习惯或者需要转为被动意义时，采用此法。有些情况下，英语被动句转译成主动句或被动句均可，取决于个人对汉语被动语态的使用习惯。

As natural habitats are destroyed, the wild plants, predatory animals, and parasites that once lived there are killed as well.

主动法：随着自然栖息地毁坏，曾经生活在那儿的野生植物、捕食动物和寄生动物也会消亡。

被动法：随着自然栖息地被毁坏，曾经生活在那儿的野生植物，捕食动物和寄生动物也会被毁灭。

汉语表达被动常用“由”“给”“受”“加以”“为……所”“使”“叫”“被”等词。

Even when use of fire became common, the relatively small amounts of smoke generated were easily and rapidly dispersed and assimilated by the atmosphere.（见 Unit 2）

译文：即使明火广泛使用，由于产生烟的数量相对较少，所以容易被大气圈快速地扩散和吸收。

She was caught in the downpour.

译文：她叫大雨淋着了。

Exercises

Ⅰ. Point out true or false.

1. Life forms have the ability to gather resources from large surroundings and make them into different forms.

2. Natural system has been overloaded from the early societies.

3. The accumulation of material wealth since the industrial civilization has led to a balanced relationship between mankind and nature.

4. Humans use natural resources to satisfy their living needs and the residues generated by the use of such resources are readily assimilated by the environment.

5. Wastewater generated by early civilizations was not so concentrated and so harmful as today, the items left behind by them were also so innocuous and their smoke were in small amount and easily dispersed in the atmosphere.

6. In 61 A. D., the waters of the Rhine and the Thames became too polluted to support the game fish and the areas harbored rats, flies and other pests.

7. The pollution overload observed in earlier times was greater than the pollution challenges faced during the industrial revolution.

8. The disposal of a toaster can contribute to solid-waste disposal problems.

9. Acquired needs are age-old needs.

10. Acquired needs are usually met by items that are readily assimilated by the environment.

Ⅱ. Answer the following questions according to the text.

1. What distinguishes humans from other life forms in terms of their ability to interact with their environment and gather resources?

2. What would be necessary to include in assessing the pollutants generated by the manufacture and sale of the toaster?

3. How did the industrial revolution contribute to a shift in human focus from natural needs to acquired needs?

4. Why is the impact of modern human populations upon the environment of major concern to the environmental engineers?

Ⅲ. Translate the following sentences.

1. In 61A.D., cooking and heating fires caused air pollution problems so severe that the roman philosopher Seneca complained of “the stink of the smoky chimneys”.

2. I’ll give you all the facts so that you can judge for yourself.

3. By the late eighteen century, the water of the Rhine and the Thames had become too polluted to support game fish.

4. The viscosity of the solution was too high to allow efficient mixing of the components.

5. The sample size was too small to draw statistically significant conclusions.

6. 我太累了，以致我一吃完晚饭就去睡觉了。

7. 显微镜的分辨率太低，不能清楚地看到亚细胞结构。

8. 我们实现了一种新的算法来优化数据处理，使计算效率提高了30%。

9. 这太平常了，不值一提。

10. 软件更新部署得非常仔细，避免了系统崩溃。

Ⅳ. Translate the following passive sentences.

1. Life-sustaining materials in gaseous, liquid, and solid forms are cycled through the biosphere, providing sustenance to all living organisms.

2. The protection of our environment must be given the highest priority because on it depends the preservation of human kind, itself.

3. The treatment facilities designed by the environmental engineer are based on the principles of self-cleaning observed in nature.（判断法）

4. The laws of conservation of mass and energy prevent the destruction of pollutants, and the engineer is bound by these limits.（被动法）

5. The potential impact of the oil spill on marine life is being assessed by researchers.

6. Trees were planted by the research team to restore the damaged ecosystem.

Ⅴ. Scan the QR code to read the article "China, Mongolia Need Better Eco-coordination", and then discuss how global warming causes dust storms, and what role China plays in the proccess of global ecological civilization governance.

Unit 3

Ecology

【重点与难点】

（1）常用构词法（bio-, aqua-, micro-等）；（2）几种表达“除……以外”的方式；（3）长句的译法。

【学习指南】

依据课程重难点，结合课程线上资源及教材的课后习题完成课前预习和课后复习，自觉应用比较学习法和主线学习法理解生态学的原理，熟悉指示生物的作用原理，掌握长句的翻译技巧，具备科技文献阅读和写作的能力。

Text 3

Bioindicator and Environment Protect

Bioindicators are used to monitor the health of an environment or ecosystem. They are any biological species or group of species whose function, population, or status can be used to determine ecosystem level or environmental integrity. An example of such a group of organisms is copepods and other small water crustaceans present in many water bodies. Such organisms are monitored for changes (chemical, physiological, or behavioral) that may indicate a problem within their ecosystem. Depending on organism selected and their use, there are three types of bioindicators: plant indicators, animal indicators and microbial indicator.

The presence or absence of certain plant or other vegetative life in an ecosystem can provide important clues about the health of the environment. Lichen is a kind of typical plant indicators in the forest. Lichens, often found on rocks and tree trunks, are organisms consisting of both fungi and algae. They respond to environmental changes in forests, including changes in forest structure, air quality, and climate. The disappearance of lichens in a forest may indicate environmental stresses, such as high levels of sulfur dioxide, sulfur-based pollutants, and nitrogen.

An increase or decrease in an animal population may indicate damage to the ecosystem caused by pollution. For example, if pollution causes the depletion of important food sources, animal species dependent upon these food sources will also be reduced in number. In addition to monitoring the size and number of certain species, other mechanisms of animal indication include monitoring the concentration of toxins in animal tissues, or monitoring the rate at which deformities arise in animal populations. ①

Microorganisms can be used as indicators of aquatic or terrestrial ecosystem health. Found in large quantities, microorganisms are easier to sample than other organisms. Some microorganisms will produce new proteins, called stress proteins, when exposed to contaminants like cadmium and benzene. These stress proteins can be used as an early warning system to detect low levels of pollution.

A biological monitor, or biomonitor, is defined as an organism that provides quantitative information on the quality of the environment around it. Therefore, a good biomonitor will indicate the presence of the pollutant and also attempt to provide additional information about the amount and intensity of the exposure.

A bioindicator is an organism or biological response that reveals the presence of the pollutants by typical symptoms or measurable responses, and is therefore more qualitative. ② These organisms (or communities of organisms) deliver information on alterations in the environment or the quantity of environmental pollutants by changing in one of the following ways: physiologically, chemically or behaviorally. The information can be deduced through the study of their content of certain elements or compounds, their morphological or cellular structure.

The use of a biomonitor is described as biological monitoring (abbr. biomonitoring) and is the use of the properties of an organism to obtain information on certain aspects of the biosphere. Biomonitoring of air pollutants can be passive or active. Passive methods observe plants growing naturally within the area of interest. Active methods detect the presence of air pollutants by placing test plants of known response and genotype into the study area. Bioaccumulative indicators are frequently regarded as biomonitors. ③ There are several types of natural biomonitors, including mosses, lichens, tree bark, bark pockets, tree rings, leaves, and fungi.

Ecological indicators are used to communicate information about ecosystems. Ecosystems are complex and ecological indicators can help describe them in simpler terms that can be understood and used by non-scientists to make management decisions. For example, the number of different beetle taxa found in a field can be used as an indicator of biodiversity.

Many different types of indicators have been developed. They can be used to reflect a variety of aspects of ecosystems, including biological, chemical and physical. Due to this diversity, the development and selection of ecological indicators is a complex process.

Using ecological indicators is a pragmatic approach since direct documentation of changes in ecosystems as related to management measure, is cost and time intensive. For example, it would be expensive and time consuming to count every bird, plant and animal in a newly restored wetland to see if the restoration was a success. Instead, a few indicator species can be monitored to determine success of the restoration. "It is difficult and often even impossible to characterize the functioning of a complex system, such as an eco-agrosystem, by means of direct measurements. The size of the system, the complexity of the interactions involved, or the difficulty and cost of the measurements needed are often crippling."

The terms ecological indicator and environmental indicator are often used interchangeably, However, ecological indicators are actually a sub-set of environmental indicators. Generally, environmental indicators provide information on pressures on the environment, environmental conditions and societal responses. Ecological indicators refer only to ecological processes.

New Words and Expressions

bioindicator *n.* 生物指示
integrity *n.* 完整性，整合性，诚实，正直
copepod *n.* 桡足动物
crustacean *n.* 甲壳纲动物
physiological *adj.* 生理学的，生理的
vegetative *adj.* 有关植物生长的，植物的，有生长力的
lichen *n.* 地衣，青苔
depletion *n.* 损耗，消耗
deformity *n.* 畸形
microbiology *n.* 微生物学
aquatic *adj.* 水生的，水产生
terrestrial *adj.* 地球的，地球上的，陆地的；*n.* 地球生物
cadmium *n.* 镉
symptom *n.* 征兆，症状

morphological	*adj.* 形态学的
genotype	*n.* 基因型
fungi	*n.* 真菌
beetle taxa	突出的分类单元
pragmatic	*adj.* 实用的
restoration	*n.* 恢复，修复
cripple	*vt.* 受损
appraisal	*n.* 评价
culminate	*vi.* 最终
methodology	*n.* 方法
sophistication	*n.* 老套，世故
cumulative	*n.* 累积的
supra-national	*adj.* 超国家的
enshrine	*vt.* 把……奉为神圣
irreversible	*adj.* 不能倒转的
rigor	*n.* 严格，严密
explicitly	*adv.* 明白地，明确地
notification	*n.* 通告，通知

Notes

① In addition to monitoring the size and number of certain species, other mechanisms of animal indication include monitoring the concentration of toxins in animal tissues, or monitoring the rate at which deformities arise in animal populations.

除了监测特定物种的规模和数量，动物指示剂还包括对动物组织毒素浓度、动物畸形率上升的监测。

② A bioindicator is an organism or biological response that reveals the presence of the pollutants by typical symptoms or measurable responses, and is therefore more qualitative.

生物指示剂是一种生物体或者生物响应，当特定的污染物存在时，该指示剂表现出典型症状或可测量的反应。因此，生物指示剂是定性的。

③ Active methods detect the presence of air pollutants by placing test plants of known response and genotype into the study area. Bioaccumulative indicators are frequently regarded as biomonitors.

主动型的方法是在研究区域放置具有已知响应和基因型的测试植物，以监测大气污染物。生物积累型指示剂常被用作生物监测。

Reading Material 3

Links between Community Ecology Theory and Ecological Restoration are on the Rise

It has been 30 years since Jordan et al. called for a "synthetic approach" to restoration, promoting the confluence of the scientific method and ecological restoration. Modern restoration programs are often carried out as factorial experiments, where explicit hypotheses are tested through analysis of quantitative data from multiple replicates and controls. Restoration sites have been referred to as "virtual playgrounds" for exploring ecology theory in real systems, echoing Bradshaw's assertion that restoration was to be the ultimate "acid test" of our ecological understanding. Though not without controversy, conducting restoration as a mode of scientific inquiry has been beneficial to fundamental ecology theory by providing opportunities to empirically validate theoretical or conceptual models in natural systems. In turn, restoration programs developed in concert with the science of ecology are widely thought to have more achievable goals, clearer methodologies, and result in more replicable and successful outcomes than restoration projects carried out following more haphazard "trial-and-error" approaches. Critics of using the scientific method in restoration argue that it fosters reductionism at the expense of using common sense to address pressing ecological issues and may bypass socioeconomic aspects of healthy, functioning ecosystems; nevertheless, the scientific method remains a popular framework for restoration activities.

With this extensive review, we have canvassed the use of various community ecology theories, concepts, and conceptually-derived tools in experimental restoration research. While some widely-cited foundational concepts such as community assembly and multispecies coexistence have maintained their presence over the past two decades, succession theory has significantly declined in use. Evolutionary dynamics are increasingly considered in the restoration of whole communities, as well as species' functional traits, highlighting conceptual foci that may yield valuable restoration insights in the future.

In light of the growing appreciation for the benefits of biodiversity for ecosystem functioning and services, restoration efforts increasingly focus on reinstating compositionally and functionally diverse biological communities in disturbed areas. In addition to the myriad difficulties in returning communities to pre-disturbance species compositions and functions (or novel ones), restoration is often tasked with keeping these attributes resilient to future environmental change. Many ecologists have cited community ecology theory as an underutilized but potentially important tool to aid in restoration design and address these challenges. Community ecology theory has particular relevance to restoration because it describes the processes that underlie the assembly, maintenance of diversity, and functioning of ecological communities which are often the focus of restoration projects.

While community ecologists should continue to seize opportunities offered by restoration activities in the form of experiments in real communities, this is not to suggest that "ecological studies conducted in restoration settings" should take priority over practical, informative restoration research. We believe that they need not, however, be mutually exclusive. Authors of fundamental ecology studies should make an effort to be more specific about restoration applications when possible, rather than paying lip service to restoration-minded reviewers. In addition, increased efforts should be made to disseminate the products of academic research to practitioners not only through high-ranking academic journals, but through wider use of extension programs, stakeholder meetings and workshops, and social media.

Similarly, practitioners should continue to explore and use elements of community ecology to their advantage whenever possible. At the very least, community ecology theory can aid in the understanding of benchmark ecological processes and functions in reference systems. Perhaps more excitingly, considering these workings in terms of established ecological tools and models that we have shown to be crossing over into the realm of restoration, such as functional traits and aspects of evolutionary theory, can yield non-traditional management solutions that may not have been apparent otherwise. We recommend that practitioners reflect on the systems from the perspectives of a variety of ecological concepts, including (but not limited to) those presented in this study, even if they do not initially seem relevant.

We acknowledge (as have other practitioners of restoration research) that restoration efforts do not need to follow purely scientific methods to advance our knowledge on the workings of natural systems. Restoration activities should ideally be informed by cultural values, economics, and policy in addition to science. Restoration is far too costly and the stakes are too high, however, to go about it haphazardly. Future restoration practitioners will have to grapple with the challenge of accommodating climate change and determining meaningful spatial and temporal scales for treatments while accounting for the variability inherent to communities, which may not yet be fully appreciated. Community ecology can provide solid, testable insights and inspiration to restoration practitioners to help address these challenges.

写作技巧

长句的译法

长句结构是英语科技文体的特点之一，它是由于英语句式具有树系结构的特点，通常围绕主干句式，大量使用限定性子句、后置定语、分词补充说明等各种复合结构而形成的。长句翻译一般可分三步走：（1）拎出主干，梳理结构；（2）从容表达，化长为短；（3）译后诵读，加工润改。

1. 分译法

对由于修辞性和限定性成分所引起的包含多层次的长句，按层次区分，将整个句子分译成几个独立的句子，顺序基本不变，保持前后的连贯。

Lichens, often found on rocks and tree trunks, are organisms consisting of both fungi and algae.（见 Unit 3）

译文：地衣通常存在于岩石和树干中，它是由真菌和藻类组成的生物体。

2. 变序法

英语习惯将传递主要信息的主句前置，后接若干含有次要信息的分句。汉语在语序顺序上是由小到大，由远及近，层层展开，最后画龙点睛，突出主题。因此，当把英语长句分译成汉语的几个小句子时，表达顺序上与英语原句有可能不一致，称为变序。

These organisms (or communities of organisms) deliver information on alterations in the environment or the quantity of environmental pollutants by changing in one of the following ways: physiologically, chemically or behaviorally.（见 Unit 3）

译文：这些生物体（或生物共同体）通过以下几种方式的变化提供环境或环境中污染物数量变化的信息：生理的、化学的和行为的变化。

3. 逆序法

英语习惯先叙“果”，后叙“因”，汉语则恰好相反。因此，遇到具有因果关系或隐含因果关系的英语长句，按汉语习惯分译过来，就会发生完全的倒序现象。

Environmental engineering has been defined as the branch of engineering that is concerned with protecting the environment from the potential and deleterious effects of human activity.（见 Unit 1）

译文：环境工程学是保护环境免受人类活动的潜在危害的学科，它已成为工程学的一个分支。

Moscow's central planners had decided to sacrifice the sea，judging that the two rivers feeding it could be put to more valuable use irrigating cotton in the central Asian desert.

译文：莫斯科的决策者觉得汇入咸海的两条河流用于灌溉中亚沙漠区的棉田更有价值，已经决定牺牲掉咸海。

Exercises

Ⅰ. Point out true or false.

1. Bioindication is a good method to monitor the health of an environment or ecosystem by plant indicator, animal indicators and microbial indicators.

2. Copepods and small water crustaceans are examples of plant indicators used to assess environmental integrity.

3. Microorganisms may produce stress proteins when they were in the toxic heavy metal soil environment.

4. Lichens, consisting of both fungi and algae, are used as animal indicators to monitor air quality in forests.

5. A biological monitor is developed from bioindication.

6. An increase in the population of a certain animal species can suggest damage to the ecosystem caused by pollution.

7. Microorganisms are harder to sample compared to other organisms, making them less suitable as bioindicators.

8. Stress proteins produced by microorganisms in response to pollution can be used to detect contamination early.

9. A biomonitor is defined as an organism that provides quantitative information about the environment.

10. Ecological indicators cost too much money and time to be used in environmental protection.

Ⅱ. Answer the following questions according to the text.

1. What is a bioindicator?

2. How many types of bioindicators are there depending on organism selected?

3. How do lichens serve as plant indicators, and what environmental changes can their presence or absence indicate?

4. What are the different ways in which microorganisms can be used as indicators of ecosystem health?

Ⅲ. Translate the following sentences.

1. The problem is particularly severe in India, where a national assessment commissioned

in 1996 found that water tables in critical farming regions were dropping at all alarming rate, jeopardizing perhaps as much as one-fourth of the country's grain harvest.

2. M is equal to the ratio obtained by dividing the prevailing molar concentrations of the oxidation products of the reaction by the prevailing molar concentrations of the reacting substances, each concentration being raised to a power equal to the coefficient of that substance in the equation representing the reaction taking in the half-cell.

3. The construction of such a satellite is now believed to be quite realizable, its realization being supported with all the achievements of contemporary science, which have brought into being not only materials capable of withstanding severe stresses involved and high temperatures developed, but new technological processes as well.

4. Collectively they underscore what is rapidly emerging as of the greatest challenges facing humanity in the decades to come: how to satisfy the thirst of a world population pushing nine billion by the year 2050, while protecting the health of the aquatic environment that sustains all terrestrial life.

5. The application of advanced remediation technologies, such as in situ chemical oxidation and enhanced anaerobic bioremediation, has demonstrated significant potential in the efficient treatment of complex contaminated sites.

6. The integration of sustainable waste management practices, such as source separation, recycling, composting, and waste-to-energy technologies, is essential for minimizing the environmental impact of solid waste disposal.

7. The establishment of environmental monitoring networks enables the continuous collection of data related to air quality, water quality, soil contamination, and biodiversity, facilitating informed decision-making in environmental management.

Ⅳ. Scan the QR code to read the article "Harmony with Nature", talk about how China's overseas infrastructure investments can maximize global biodiversity conservation goals while ensuring local development is sustainable.

Unit 4
Environment Management

【重点与难点】

（1）环境保护和管理方面常用的专有名词和缩略词；（2）常用构词法；（3）对“少数”的几种表达方式；（4）翻译中词义的改换和词序的调整。

【学习指南】

依据课程重难点，结合课程线上资源及教材的课后习题完成课前预习和课后复习，自觉应用科学发展观指导环境影响评价，掌握词义改换和词序调整的技巧。

Text 4

Principle of EIA Administration and Practice

The architects of NEPA intended the environmental impact statement to be the “action-forcing” mechanism, which would change the way government decisions were made in the USA. ① However, they probably did not foresee the extent to which EIA would be adopted internationally, culminating in Principle 17 of the Rio Declaration on Environment and Development. ② Today, EIA is applied in more than 100 countries, and by all development banks and most international aid agencies. EIA has also evolved significantly, driven by improvements in law, procedure and methodology. In all countries, more strategic, sustainability-based approaches are still at a relatively early stage.

(1) 1970—1975: introduction and early development. Mandate and foundations of EIA established in the USA; then adopted by a few other countries (e.g., Australia, Canada, New Zealand); basic concept, procedure and methodology still apply.

(2) Mid 1970s to early 1980s: increasing scope and sophistication. More advanced techniques (e.g., risk assessment); guidance on process implementation (e.g., screening and

scoping); social impacts considered; public inquiries and reviews drive innovations in leading countries; take up of EIA still limited but includes developing countries (e.g., Thailand and the Philippines).

(3) Early 1980s to early 1990s: process strengthening and integration. Review of EIA practice and experience; scientific and institutional frameworks of EIA updated; coordination of EIA with other processes (e.g., project appraisal, land use planning); ecosystem level changes and cumulative effects begin to be addressed; attention given to monitoring and other follow-up mechanisms. Many more countries adopt EIA; the European Community and the World Bank respectively establish supra-national and international lending requirements.

(4) Early 1990s to date: strategic and sustainable orientation. EIA aspects enshrined in international agreements; marked increase in international training, capacity & building and networking activities; development of Strategic Environmental Assessment (SEA) [③] of policies and plans; inclusion of sustainability concepts and criteria in EIA and SEA practice; EIA applied in all Organization for Economic Cooperation and Development (OECD) countries and large number of developing and transitional countries.

To date, EIA has been applied primarily at the project-level. This "first generation" process is now complemented by SEA of policies, plans and programmes, and both EIA and SEA are being adapted to bring a greater measure of "sustainability assurance" to development decision making.[④] These trends have brought new perspectives on what constitutes EIA good practice and effective performance.

Recently, a number of reviews of these issues have been undertaken, including the International Study of the Effectiveness of Environmental Assessment. It described basic and operational principles for the main steps and activities undertaken in the EIA process. The International Association for Impact Assessment (IAIA) and the Institute of Environmental Management and Assessment (IEMA) have drawn on these to prepare a statement of EIA best practice for reference and use by their members. The effectiveness study identified three core values on which the EIA process is based: (1) Integrity—the EIA process should meet internationally accepted requirements and standards of practice. (2) Utility—the EIA process should provide the information which is sufficient and relevant for decision-making. (3) Sustainability—the EIA process should result in the implementation of environmental safeguards which are sufficient to mitigate serious adverse effects and avoid irreversible loss of resource and ecosystem functions.

New Words and Expressions

EIA	Environmental Impact Assessment 环境影响评价
proposal	*n.* 提案，提议，建议，申请，求婚
appraisal	*n.* 评价
architect	*n.* 建筑师，设计师，本文引申为（法律）编撰者
foresee	*vt.* 预知，预料
adopt	*vt.* 过继，采纳，接受，通过，批准
culminate	*vi.* 最终
methodology	*n.* 方法
strategic	*adj.* (strategical) 战略上的，策略上的
mandate	*n.* 授权，正式命令；*vt.* 将（某地）委托某国管理，授权；*vi.* 强制执行，委托办理
sophistication	*n.* 老套，世故，复杂性
implementation	*n.* 执行，完成，贯彻，成就
coordination	*n.* 协调，和谐
project appraisal	项目评估
land use planning	土地（利用）规划土地规划
cumulative	*adj.* 累积的
follow-up	*adj.* 作为重复的，继续的，接着的
supra-national	*adj.* 超国家的
orientation	*n.* 方向，目标
enshrine	*vt.* 把……奉为神圣
criterion	*n.*（批评、判断的）标准，准则，尺度，复数形式 criteria
OECD	经济合作与发展组织（经合组织）
transitional	*adj.* 过渡的，变调的
undertake	*vt.* 着手，承揽，担任
irreversible	*adj.* 不能倒转的

Notes

① The architects of NEPA intended the environmental impact statement to be the “action-forcing” mechanism, which would change the way government decisions were made in the USA.

国家环境保护法（NEPA）的制定者企图使环境影响报告具有“强制性”，这将改变美国政府的决策方式。

② the Rio Declaration on Environment and Development

环境与发展里约宣言。共27条原则，其中第17条指出：应对可能会对环境产生重大不利影响的活动和要由一个有关国家机构作决定的活动作环境影响评估，作为一个国家手段。

③ Strategic Environmental Assessment (SEA)

战略环境评价。SEA是对政府政策、规划及计划的环境影响评价。

④ To date, EIA has been applied primarily at the project-level. This “first generation” process is now complemented by SEA of policies, plans and programmes, and both EIA and SEA are being adapted to bring a greater measure of “sustainability assurance” to development decision making.

目前，环境影响评价（EIA）主要应用在建设项目水平。针对政策、规划和方案的战略环境评价（SEA）是“第一代”的环评程序的补充。EIA和SEA都致力于提供更充分的评价方法，以考量项目开发的决策是否能够“确保可持续性”。

Reading Material 4

Integrating Climate Change in Environmental Impact Assessment: A Review of Requirements across 19 EIA Regimes

Overall, the results show that climate change integration in EIAs is a new and necessary challenge for the practice internationally. Through assessment of levels of integration (radical, partial) and non-integration, this research has identified that within 19 regimes reviewed, impacts of climate change, mitigation and adaptation are explicitly integrated in a variety of EIA guidelines, as well as in various stages of an EIA. In contrast, there remains multiple regimes with no requirement for climate change consideration in their EIA regulations. This is in part likely because EIA regulations were established in an era where climate change was not widely considered and integrated, and the focus of EIAs was to achieve development approval. Nonetheless, new regulations and guidelines, such as the EU Directive have promoted integration of climate change in EIA, have provided opportunities for such iterative and substantive amendments.

Climate change integration is not required in most jurisdictions. For those that do have climate change requirements, there is a wide diversity of integration requirements. For example, in the USA, practitioners are not legally required to address climate change in EIAs, however, guidelines from governments including CEQ and private organizations including IEMA advise them to integrate GHG emissions and associated impacts. This study identified that only the EU and UK have a specific legal requirement that EIAs should integrate climate change. Many other regimes in this study provide guidance for practitioners conducting an EIA but do not require explicit integration in their regulations.

Regulations and guidelines on management and monitoring plans could better require assessment of GHG emissions and project level adaptation to climate change, particularly for project alternatives considerations. Appropriate considerations of project alternatives can serve as mitigation and adaptation measures. However, in most regimes, there was no consideration of adaptation, which fundamentally undermines meaningful climate integration.

This study indicates that lack of climate change integration in EIAs is mainly caused by the absence of obligatory requirements, guidance, and experience in integrating climate change into assessments, along with the linear process of EIAs. Including the obligation to integrate climate change in the EIA regulations is crucial. However, more measures are likely to be necessary to ensure that they are appropriately implemented. For instance: practical guidance, training and motivational programmes for practitioners, and awareness creation amongst all EIA stakeholders.

There are several challenges when integrating climate change considerations into the EIA. The following recommendations are made in relation to the identified challenges and scope potential foci of future research directions, regulatory development, and practice guidelines.

Legislative reform can provide detailed provisions within regulations on how to address climate change, and a guiding framework for such explicit integration is required. These types of reforms can target higher levels of integration where EIAs can be "climate change based" or "climate testing" or set minimum standards of integration above threshold requirements for "partial integration".

A range of broad recommendations can be drawn from the strengths and weaknesses of the 19 EIA regimes reviewed here. Climate change should be integrated holistically as a consolidative part of the EIA and not constrained to only one or two steps in the process. Explicit requirement for consideration of climate change adaptation and integration of climate change mitigation and adaptation measures is needed to balance current emphases on mitigation. Establishing and strengthening post-decision monitoring of climate change impacts is crucial. Environmental Management Plans need periodic monitoring by frequently assessing and updating knowledge concerning environmental changes, climate trends and baselines. Regulations and guidelines need to have broad sectoral coverage and not exclude certain sectors, particularly those with greatest relevance to climate change including energy production, transport, mining, infrastructure, and housing. Greater alignment between integration of climate change considerations within EIAs and those in Strategic Environmental Assessments is also desirable. The development of a high-level screening tool would help identify likely climate change aspects, specifically, one able to be calibrated to the sectoral and geographical local climate change information needs of EIA practitioners.

Future research can extend these findings through conducting systematic review of all EIA regimes, including those in languages other than those included in this study. It can also test how well metrics of integration scope and procedure used here can be applied in other contexts. A particularly urgent research agenda arising from these results concerns the standardization of integration of climate change considerations in EIA in both regulations and guidelines. Across multiple contexts this will require deep contextual reflection regarding what works within specific EIA regimes and where potential progress can be made. Concerning timing, a significant investment of research is needed to align both the practice of EIA and pace of regulatory developments with the urgency of the climate crisis making this a particularly urgent research agenda.

翻译中词义的改换和词序的调整

英汉两种语言在词汇、结构、修辞和文化背景上有很多差异，正是因为这些差异，改换在翻译中是不可避免的。为了达到译文与原文等值的效果，在翻译时常采用词义引申、词性转换和词序调整的方法。

1. 词义引申

英语词的含义范围比较宽，并且词义对上下文的依赖性比较大，在基本词义的范围内可随语境而变化。

（1）词义转译

英语单词普遍存在一词多义、一词多类的现象，需要根据上下文、词类及词在句中的搭配来判断和确定某个词在特定场合所应具有的词义。

EIA is one of a number of policy tools that are used to evaluate project proposals.（见Unit 4）

译文：环境影响评价是一种政策工具，它用于项目可行性的评价。

It's incumbent on structural engineers to design buildings in harmony with environmentally sound requirement.

译文：设计对环境友好的建筑物是建筑工程师义不容辞的事。

（2）词义的具体化和抽象化

英汉两种语言在表达事物共性和个性的逻辑习惯上差异较大，汉语习惯具体表达的场合在英语中往往表达为概括性，而英语中又常使用具体意义的词来表示一种属性、一类事物或一个概念。在英译汉时，需要根据汉语的习惯，把英语原文中意义笼统或抽象的词表达为意义较明确或具体的词，或者把英语中使用具体意义的词转化为汉语中笼统和抽象的词。

Agriculture has been identified as the major contributor to non-point source pollution in watershed .

译文：农业已确认为流域环境非点源污染的主要作用因素。

Considerable press coverage has been given to contamination of water by mercury.

译文：相当多的新闻媒体报道了由水银造成的水污染。

2. 词性转换

英汉两种语言词的使用范围和表达方式不尽相同。在大多数情况下，翻译句子时不能逐词对译，随着词类的变化，句子的结构也产生了变化。

（1）译为动词

英语主谓机制突出，一个句子中往往动词少，名词多，尤其是抽象名词用得多，常靠词形变化来表达意思；汉语没有词形变化，重动态描写，所以汉语动词用得多，表达意思时往往借助动词，按时间及逻辑循序层层交代。在英译汉时，英语中许多意思都可以用汉语的动词来表达。

The protection of our environment must be given the highest priority because it depends on the preservation of human beings, itself.（见 Unit 2）

译文：我们应该把环境保护放在首位，只有这样才能保护人类自己。（名词转译成动词）

The use of fossil fuels for heating and cooling, for transportation, for industry, and for energy conversion, and the incineration of the various forms of industrial, municipal, and private waste all contribute to the pollution of the atmosphere.

译文：采用化石燃料用于供暖和制冷、服务交通运输、满足工业、进行能量转换以及焚烧各种工业、市政和个人排放的废物等活动是造成大气圈污染的主要因素。（介词转译成动词）

Transportation was responsible for 56 % (by weight) of all the contaminants emitted to atmosphere in 1977, as compared to 55 % in 1980.

译文：与 1980 年的 55%比较，1977 年流动运输排放的污染物占排放到大气圈的全部污染物的 56 %（按重量）。（形容词转译成动词）

If both sources are on at the same time, their acoustic pressures cancel. (on 为副词，进入与某物相接触的位置或状态)

译文：若两个声源同时相遇，它们的声压相抵消。(某些副词转译成汉语的动词)

(2) 译为名词

EIA has also evolved significantly, driven by improvements in law, procedure and methodology.（见 Unit 2）

译文：通过对法律条款、程序和方法的改进，环境影响评价也取得了显著进展。(动词转译成名词)

Glass is much more soluble than quartz.

译文：玻璃的溶解度比石英大得多。(形容词转译成名词)

某些表示事物特征的形容词作表语时可转译成名词，加上“性”“度”“体”等。

(3) 其他词类转译

The modern world is experiencing rapid development of science and technology.

译文：当今世界科学技术正经历迅速的发展。(形容词与副词的互相转译)

Earthquakes are closely related to faulting.

译文：地震与断裂运动有密切的关系。

3. 词序调整

英汉两种语言的语序差异主要表现在修饰成分词序上的差异，英语修饰语位置比较灵活，汉语修饰语位置比较固定，同时两种语言由于句法结构和表达习惯不同也可引起差异。因此，在英汉互译过程中，需要按照语言的词序进行必要的调整。

(1) 句法结构和表达习惯不同引起的差异

英语中，当涉及一系列的措辞或词组的排列时，总是按比喻力度的大小，从弱到强排序，汉语则比较随意。例如：

The containers used for disposal of harmful wastes，toxic chemicals and radioactive materials must isolate those materials from the environment in order to prevent the onset of the natural, but highly undesirable, processes of dilution and dispersion.（见 Unit 4）

译文：采用容器封装使放射性、有毒和有害的废物与环境隔离，以防止极不希望的自然扩散和稀释作用的发生。

针对项目的罗列，英语习惯先说总数，再详列项目，而汉语则相反。

With the gained fund, this supervision station has in the last two years bought over 50 instruments, including 10 noise gauges, 5 spectrographs, 30 automatic COD testing, in addition to flow meters and anemometers.

译文：这个监理站用争取来的资金，近两年里购买了 10 部噪声计、5 台光谱仪、30 台 COD 自动测试仪，以及流量计、风速计等共 50 多套设备。

(2) 修饰成分词序上的差异

两个以上后置定语一般不是平行关系，它们并不独立地修饰中心词，而是最后一个限定它前面的一个定语，两个定语一起再限定前面一个定语。译成汉语时一般要按连锁关系的各个层次倒译。例如：

An understanding of natural and engineered purification processes requires an understanding of the biological and chemical reactions involved in these processes.（见 Unit 1）

译文：了解自然和工程的净化作用需要了解包含在这些过程中的生物和化学反应。

英语中一个中心词既有前置定语又有后置定语，译成汉语时一般将后置定语放在前置定语之前。例如：

The delicate balance of our biosphere has been disturbed and the state is a direct consequence of our having ignored the limits of the earth's ability to overcome heavy pollution loads.

译文：生物圈脆弱的平衡已经破坏，这种状况是一直忽视地球克服严重污染的有限能力的直接结果。

（3）状语位置的变换

状语有三种形式：句中式、句首式和句尾式。汉语多取句中式，即多位于主谓之间，或取句首式，即位于句首，基本上无句尾式。英语多用句尾式，也用句中式和句首式。对于常见的因果关系，英语习惯于“先果后因”的叙述，汉语习惯于“先因后果”的叙述。

The rockets are very powerful, armed with high-explosive warheads.

译文：由于装上了高爆弹头，这种火箭的威力很大。

时间、地点和方式副词的词序。当句子中同时出现时间、方式和地点副词时，汉语的语序习惯是“时间（time）—地点（place）—方式（manner）”，英语中的自然语序是“方式（manner）—地点（place）—时间（time）”。

We ate to our hearts' content (manner) at her home (place) last Sunday (time).

译文：我们上星期天（time）在她家（place）尽情地（manner）吃了一顿。

表示时间和地点，汉语是先大后小，而英语则是先小后大。例如：

His address is 111 of Wenjiao Road, Nanchang, China .

译文：他的地址是中国南昌文教路 111 号。

My daughter was bored at 2: 30 pm. on August 12, 2010.

译文：我女儿生于 2010 年 8 月 12 日下午 2 点 30 分。

Exercises

Ⅰ. Point out true or false.

1. EIA is used to evaluate the effect of the planning and existed project on the

environment, and is established first in the USA.

2. EIA is widely applied in many countries including China, and many new methods such as public inquiries are used in EIA.

3. SEA is a system of incorporating environmental considerations into policies, plans and programs.

4. NEPA architects intended the environmental impact statement to have a limited impact on government decisions in the USA.

5. The development of EIA has not been influenced by improvements in law, procedure, or methodology.

6. Strategic, sustainability-based approaches to EIA have been widely adopted in all countries.

7. EIA was primarily applied at the project-level until the introduction of SEA.

8. EIA process is not concerned with addressing cumulative effects and ecosystem level changes.

9. The Rio Declaration on Environment and Development played a significant role in the international adoption of EIA.

10. EIA process is based on three principles (integrity, utility and consistency) to realize the sustainable development.

Ⅱ. Answer the following questions according to the text.

1. What core values is EIA process based on?

2. What is SEA when compared with EIA.

3. How has EIA evolved over time, and what factors have driven its evolution?

4. What are the three core values identified in the effectiveness study as the basis for the EIA process?

Ⅲ. Translate the following sentences.

1. In the view of environmental economics, economically sound projects do not in themselves constitute practical decision-making.

2. The problem in manufacture is the control of contamination and foreign materials.

3. Meeting the acquired needs（习得需求）of modern societies generates more residuals than meeting natural needs.

4. Radiation features a longterm effects.

5. In fission processes the fragments are very radioactive.

6. Environment science aims to protect the environment from the potential, deleterious effects of human activity.

7. He is the distinguished Canadian scientist.

Ⅳ. Scan the QR code to read the article "River Cleanup Results in Numerous Benefits" , and then discuss what is the Scientific Outlook on development, how does it guide people's work in evironmental protection?

Unit 5

Environmental Monitor

【重点与难点】

（1）环境监测领域常用的术语；（2）常用构词法；（3）科技文献中图表的表达方式；（4）科技文献中举例、倍数以及其他常用结构的互译。

【学习指南】

依据课程重难点，结合课程线上资源及教材的课后习题完成课前预习和课后复习，掌握科技文献互译的典型句型，结合环境监测要求与规范，自觉成为求真务实、诚实守信、吃苦耐劳的预环境工程师。

Text 5

Biological Monitoring in Terrestrial Systems

Pollution on land can arise from many sources such as industrial discharges, municipal refuse, mining activities, disposal of sewage sludge, motor vehicles and also deposition from the atmosphere and water. We are thus concerned with a wide range of substances. Many of these are difficult to detect but others, such as heavy metals, can be monitored by using biological means. As with aquatic systems, monitoring on land has often centered around land usage, e.g. cereal or crop growth, amenity or wild life preservation.

Many terrestrial organisms are affected by air and land pollutants. For example, a tree will be affected by soil-based pollutants through its roots and air-based pollutants through its leaves. One must always, therefore, be careful in one's interpretation of the concentrations of pollutants presented in an organism in these situations. Chlorosis (yellowing) of the leaves in plants is an example which could be due either to soil nutrient deficiency or SO_2 contamination of the atmosphere.

Plants and animals can be used to monitor pollution on land. Vascular, or higher plants often require specialist knowledge to recognize pathological changes due to pollution, however. Perhaps a better way to use them is to take advantage of their capacity to absorb contaminants, as will be seen later. In recent years there has been an increasing number of surveys using small animals for monitoring pollution, e.g. metals, especially around industrial sources and roads.① Many species of small mammals are common in rural areas in many countries. These can often be easily caught by means of long worth traps and their tissues can be analyzed. This technique is attractive in that one is using an animal not too different to man in its physiology or in its position in a food chain especially when compared with plants or invertebrates. This approach has been used by a number of authors with varying success at sites for comparison. Examples of results obtained are given in Table 5-1.

These results especially from USA, show that small mammals do exhibit increased body burdens of contamination near to pollution and that this changes with source intensity and distance.② To be of any use in monitoring, the response of the indicator should be quite definite and, if possible, graded. Small mammals can show these types of response but various authors have found considerable variation between individuals at the same site, which can make interpretation of small differences in pollution loadings difficult.

Table 5-1 Concentration of Pollutants in Small Mammals Collected from Various Locations

Location	Ref	Source	Contaminant/$mg \cdot kg^{-1}$			
			Lead	Mercury	Copper	DDT
Mainroad	66	Microtus[a]	3.31	—	—	—
		Apodemus[a]	1.37	—	—	—
		Vegetation[a]	226.00	—	—	—
Field	69	Apodemus[a]				
		Whole	—	0.85	13.3	1.02
		Liver	—	1.35	5.0	—
Roadside						
Light traffic	67	Peromycus[b]	0.70	—	—	—
Heavy traffic			4.60			
Light traffic		Soil	~900			
Heavy traffic		Soil	~1100			

Note: a: wet weight; b: dry weight.

For straightforward monitoring purposes on land evidence suggests that herbage may exhibit a greater and better response to pollution. This is particularly true for metals. This is shown well with collections of plants at varying distances from roads in order to assess the extent of lead pollution. One has to be careful to collect only the same species of plant at each site as there can be variation from one species to another. Examples of such a survey are given in Table 5-2.

Table5-2 Concentrations of Lead in Plant Leaves and Soil Unit: $mg \cdot kg^{-1}$

Site	Vegetation type			
	Dock	Rhubarb	Elder	Soil
Near battery factory	1474.0	1807.0	1219.0	4570.0
Busy road				
At road verge	83.0	—	72.0	625.0
10m distance	60.4	25.0	53.0	340.0
20m distance	42.0	19.0	36.0	120.0

These results show that both vegetation and soils respond to environmental contamination in a graded way. Pollutant concentrations are higher in the soil samples in the case shown although this is not always so. The concentrations in the soil do not necessarily represent what is biologically available as the lead, in this case, may be bound chemically in an insoluble form. ③

It has been found that for many pollutants broad leafed plants give a better indication than narrow leafed ones. Also, if one is interested only in the pollutant that is absorbed into the plant, it is a good idea to wash the outside to remove surface contamination as this represents aerial pollution rather than that truly in the soil.

New Words and Expressions

terrestrial	*adj*. 陆地的，长于陆地的，地球的，世界的
municipal	*adj*. 市政的，地方政府的
sewage	*n*. (下水道里的)污物
sludge	*n*. 污水，下水道里的污泥
aquatic	*adj*. 水生的；*n*. 水生植物
cereal	*n*. 作食物用的任何谷物，谷子，谷物；*adj*. 谷类的
amenity	*n*. 舒适，适宜，愉快，乐事
chlorosis	*n*. 萎黄病，变色病
yellow	*vt*. 变黄；*n*. 黄色；*adj*. 黄的
vascular	*adj*. 管的，血管的，含管的，含血管的，导管的
pathological	*adj*. 病理的，病害的

longworth *adj*. 长效的
trap *n*. 陷阱，捕捉机，存水湾；*v*. 设陷阱，使陷于
invertebrate *n*. 无脊椎的，无骨气的，无脊椎动物
woodland *n*. 林地，森林，树林
vegetation *n*. 植物，草木
indicator *n*. 指示者，指示物，指示剂，指示器，火车指示牌
individual *adj*. 特别的，独特的；*n*. 个人
straightforward *adj*. 诚实的，坦白的，易懂的，易做的
herbage *n*. 草本植物，牧草
mammal *n*. 哺乳动物
dock *n*. 月台，站台，飞机库，酸模，羊蹄草
rhubarb *n*. 大黄
elder *n*. 接骨木
grade *n*. 阶级，品位，等级，班级，年级，分数，成绩
aerial *adj*. 存在空气中的，空中的，航空的，无形的，虚幻的
verge *n*. 边缘

Notes

① In recent years there has been an increasing number of surveys using small animals for monitoring pollution, e.g. metals, especially around industrial sources and roads.

近年来，采用小动物监测污染物（例如重金属，特别是工业企业和公路附近的重金属）的调查数量日益增加。

其中，increasing 译为"日益增加"。

② These results especially from USA, show that small mammals do exhibit increased body burdens of contamination near to pollution and that this changes with source intensity and distance.

尤其是美国的研究结果表明，污染附近的小哺乳动物体内污染负荷增加，污染负荷随着污染的强度和距离发生变化。

③ The concentrations in the soil do not necessarily represent what is biologically available as the lead, in this case, may be bound chemically in an insoluble form.

实验结果表明，土壤中铅的含量不能代表植物可利用态的铅，铅会以不可溶的形式结合在土壤中。

土壤中重金属的形态可分为水溶态、可交换态、碳酸盐结合态、有机结合态、铁锰结合态等形态，其中水溶态和可交换态为植物直接可利用态，含量较低。

Reading Material 5

A Review of Remote Sensing for Environmental Monitoring in China

Natural environment is a general term for water resources, land resources, biological resources and climate resources, which affects human beings' survival and development. It is closely related to the sustainable development of society and economy. Since the Industrial Revolution, the intensity of the human use of natural resources has continuously increased with the rapid growth of populations, and in particular, the overexploitation of resources leading to the deterioration of the environment, has become greater and greater. For example, the consumption of fossil fuels and the deforestation of forest resources accelerate global climate warming, and further lead to the extinction of biological species at an unprecedented rate. When the environmental load exceeds the limit that an ecosystem can bear, the ecosystem would gradually weaken and be exhausted. Therefore, natural environment monitoring is of great significance for the environmental resource protection and management, and the survival and development of human beings.

The commonly used ground-based monitoring is limited by regions and only suitable for point-based environmental monitoring in a small region. This method is time-consuming, costly and labor intensive. Due to the characteristics of large-scale and dynamic observation, remote sensing sensors are able to quickly acquire the wide spectral information of regional and even global targets, such that various ecological indicators can be obtained by data modeling and retrieval. As such, remote sensing has been an indispensable approach to ecological monitoring gradually, especially at a large or global scale.

As a developing country, China is experiencing a significant change in natural environment, and remote sensing is playing an important role in environmental monitoring and protection. Many agencies carried out a large amount of work with remote sensing to monitor ecological status and change. As an example, Ministry of Ecology and Environment Center for Satellite Application on Ecology and Environment, has conducted a great amount of work by using international satellite data such as Landsat, MODIS and China-launched satellite data such as HJ-1A, Gaofen-1, Gaofen-2. They have become operational in many services including ecological index retrieval, human activity monitoring in protected areas, and environmental damage assessment and so on, which provides strong support for national environment management and decision-making. This is very useful for the official inspection on ecological and environmental protection, especially for the Green Shield Action established by seven state ministries. Two typical cases are environment degradation in the Qilian Mountain National Nature Reserve in Gansu Province and the illegal villa construction at the northern foot of the Qinling

Mountain in Shaanxi Province. In addition, many studies and literatures documented the progress in many fields, such as ecological index retrieval, urban and rural area monitoring, and nature reserves monitoring. However, few studies have systematically and comprehensively summarized the progress in environmental monitoring by remote sensing in China in recent decades.

This paper aims at reviewing the satellite resources, institutions and policies for environmental monitoring in China, and the advances in research and application of remote sensing from five aspects: ecological index retrieval, environmental monitoring in protected areas, rural areas, urban areas and mining areas (atmosphere and water environment monitoring will be introduced separately). This paper also points out major challenges existing in the current stage and puts forward the development trend and future directions, which is useful to direct the research and application of environmental monitoring and protection in the new era.

Although China has made great achievements in improving the environment, it still now faces serious environmental pressure. The environmental degradation occurs frequently due to economically motivated activities. To maintain environmental safety, lucid waters and lush mountains, it is very necessary for China to strengthen the protection and supervision of the environment with remote sensing technology, especially for the China Ecological Conservation Red Line (ECRL), implying the areas which have especially important ecological functions and must be carefully protected. Currently, the Ministry of Ecology and Environment of China, is carrying out remote sensing supervision for ECRL and national parks.

Nevertheless, in view of the existing problems, the following directions are issued so as to promote the technological development and meet the requirements of national environment management.

(1) Promoting multi-source data fusion. Multi-source remote sensing data can make up for the shortcomings of the single type of data on temporal resolution, spatial resolution or spectral resolution and provide complementary information. Therefore, how to estimate ecological index using multi-source data fusion including remote sensing data, ecological data, environmental data, meteorological data, social and economic data, is an important direction.

(2) Improving the retrieval accuracy of remote sensing parameters. Considering the uncertainty of the remote sensing retrieval process, the accuracy of ecological parameter retrieval is improved by optimizing models and developing land surface data assimilation systems. Presently, the ecological index retrieval is mainly based on the relationship

between remote sensing data and ground-measured data. Verification relies on measured data, but there are also errors in the process of obtaining in-situ data. Thus, the accurate evaluation of model retrieval and how to reduce the dependence degree of retrieval models on ground-measured data are research directions in the future.

(3) Modelling of scale effect. The scale effect is the main problem that remote sensing of the environment needs to face. The accuracy of remote sensing retrieval is related to the spatial and temporal resolution of the dataset used. Therefore, it is necessary to select an appropriate spatiotemporal scale according to the requirements, and further select an available remote sensing dataset.

(4) Improving the level of production automation. Besides the accuracy of products, the state and the public also have high requirements in the timeliness of products in many application scenarios, which directly impacts the quality of applications. A focus of future work will be to increase the automation level of remote sensing products through process optimization, customized function development and aggregating computational resources.

(5) Enhancing the ability of forecasting and comprehensive analysis, shifting remote sensing of the environment from monitoring to a synthetic ability with monitoring and forecasting, and strengthening the comprehensive analysis of results are important directions for future work.

(6) Promoting open access to satellite datasets. At present, China's satellite data resources are mainly concentrated in the China Center for Resources Satellite Data and Application, and a few ministries and agencies. The colleges, research institutions and the public have a relatively poor accessibility to domestic satellite data. It is necessary to formulate relevant policies and protocols to increase the openness of satellite data, to promote applications, and to enhance information dissemination.

(7) Implementing ECRL supervision by remote sensing. There is an urgent need to supervise the ECRL in constructing the national ecological safety and ecological civilization, which can effectively restrain ecological damage in China and provide a reference for other countries.

写作技巧

科技文献中部分常用结构的互译

1. 举例

文章以什么形式进行展开取决于论文本身的特点。对于一个段落来说它的内容由主题句（topic sentence）的导向决定。有时候，举例是说明主题最快捷、最直接的方法。

（1）举例的标记

一般来说，作者在举例之前会给读者发出一个“信号”，说明下面要举例了。这个信号就是“for example”“for instance”或其他标记。如果要重复强调，必须使用不同的方法说明。例如：

Many terrestrial organisms are affected by air and land pollutants. **For example**, a tree will be affected by soil-based pollutants through its roots and air-based pollutants through its leaves.（见 Unit 5）

当然，在阅读中，也会发现有时作者举例前没有给出“信号”，这是因为作者觉得读者能识别他所用的例子。

Many people collect things as a hobby. They may collect stamps, old coins, matchbox covers, dolls, phonograph records, or autographs of famous people.

（2）举例的类型

- 以“when”引导的排比句为信号

Paul always participates in class activities. **When** his history instructor gives a review for an examination, he eagerly raises his hand so he may answer questions. And **when** his sociology professor conducts a class discussion, he always contributes useful material.

- 用罗列举例法，一般有明确的提示，通常为“for example”的变形方式。例如：“there are many examples of”“following examples demonstrate”和“ as the following examples demonstrate”等，例如：

Symbolism is a major means often used in literature, painting and film to make the works more vivid and meaningful. **There are many examples of** symbolism in Beijing opera as well.

The following examples demonstrate how orders are matched through closing auction and details of the algorithm applied for determining the Indicative Equilibrium Price (IEP). However, the examples shown are not exhaustive.

Manual labor is one of the principal development resources in any industrializing country, **as the following examples demonstrate**.

需要说明的是，当主题句已明确说明“如下例所示/as the following example demonstrate”，展开部分就不必再用“for example”这样的标记了，只要用罗列连接词语就行了，如：To begin with, Secondly 和 Finally 等。当主题句没有明确指出举例说明时，展开句要一一说明，如 For example, As another example, As still another instance 等。

- 名词的罗列

科技文献不仅在陈述主题句时需要举例，而且在单句的表达中也常常出现名词的罗列现象，现分别举例如下：

Six vegetables **namely** tomato, cucumber, French bean, radish, spinash and Chinese cabbage, were irrigated with wastewater.

利用废水灌溉了番茄、黄瓜、菜豆、萝卜、菠菜和白菜等六种蔬菜。

Any material, which may be gas, liquid or solid is made up of atom./Any material, such as gas, liquid or solid, is made up of atoms.

任何物质，不论是气体、液体还是固体，都是由原子组成的。

Some branches of mathematics such as **probability theory and group theory** are supplied in an **increasing** range of activities.

数学的一些分支，如**概率论、群论**，应用范围日**益**扩大。

2. “是”“有”“呈”的译法

（1）“是”的译法

汉语中“是”的作用：判断主语是什么东西或属何种类（be 或其他动词）；描绘或说明主语的性质（be 或其他动词）；是……的（it is …that or who 或省略不译）。现分别举例如下：

- be 或其他动词

全世界科学家和科学机构之间的相互合作日益密切，是当代科学最显著的特征之一。

译文：One of the most characteristics of the modern science **is** the closer co-operation between scientists and scientific institutions all over the world.

这个方法实际上**是抑制**废水的产生。(只译后面的动词)

译文：This method actually **prevents** the production of wastewater.

文中讨论了许多问题，其中之一**是**关于这个理论的。

译文：Many problems are discussed in this paper, one of them **is** about this theory.

- 是……的（it is …that or who 或省略不译）

地球上包括人在内的生物形式正**是**生活在生物圈**的**。

It is within the biosphere **that** the life forms of earth, including humans, live.（见 Unit 1）

（2）“有”的译法

汉语中的“有”可译成以下三种形式：译成 There be, have 结构；译成其他动词；译成介词结构，如 with。现分别举例如下：

- There be, have

Humans alone **have** the ability to gather resources from beyond their immediate surroundings and process those resources into different, more versatile forms.（见 Unit 2）

译文：只有人类**有**能力在周边环境以外的地方获取资源，并把它们加工成不同的、更丰富的形式。

- 其他动词

室内装有空调设备。

译文：The air-condition was equipped in the room.

放射学的成果现在有着许许多多不同的用途。

译文：The results of radioactivities **are** now **applied** in a great number of different ways.

水银气压计（barometer）有一个超过76cm、一端封口的充满水银的细玻璃管组成。

译文：A mercury barometer **consists of** a narrow glass tube, over 76cm long, which is closed at one end and filled with mercury.

- 介词结构

有一些国家人口过剩，有一些国家人口下降。

译文：**In some countries**, overpopulation developed, while in others a decline in population occurred.

铆钉是一端有头的一种金属销钉。

译文：The rivet is a metal pin **with** a head at one end.

（3）“呈”的译法

“呈”常以“呈某种形式、呈某颜色、呈某种状态、呈正（负）相关”等表达形式出现在科技文献中。现举例如下：

- 呈某种形式

诱导的愈伤组织呈米黄色或黄绿色，结构坚实。

译文：The callus induced is yellowish or yellowish green and compact in texture.

- 呈某种形状

锤尖通常呈圆形或楔形。

译文：The peen of the hammer is often rounded or wedge-**shaped**（wedge **in shape** or wedge **shape**）.

- 呈某种状态

呈气态或蒸气态的水叫蒸汽。它是由沸腾的水产生的。

译文：Water in the form of gas or vapor (Water as a gas or vapor) is called steam. It is produced by boiling water.

- 呈正（负）相关

温度与声速呈显著正相关。

译文：A significant positive correlation exists between temperature and speed of sound.

需要说明的是，在表达两者相关性时也可以用“there be”句型，即 there is a positive(negative) relationship between A and B。

目前，我国村镇两级居住社区的人均生活垃圾产生量，分别为0.5～1.0kg·(人·d)$^{-1}$

和 0.4～0.9kg·(人·d)$^{-1}$，产生量与地域经济发展水平正相关。

The current production rates of RDW (rural domestic waste) in residential areas of vilages and towns were 0.5-1.0kg·(capita·d)$^{-1}$ and 0.4-0.9kg. (capita·d) $^{-1}$ respectively, which were positively correlated with the local economic leve1.

3. 对比、比较、倍数、百分数的表达

（1）对比表达

常用的连词和词组有 in contrast, compared with, in comparison with, on the other hand, while。

In contrast to the robust increases of 2.1 percent per year between 1950 and 1990, the rise between 1990 and 1995 averaged only 1 percent a year.（见 Unit 10）

译文：1950—1990 年间（粮食产量）每年按 2.1%比例快速增长，与之相比，1991—1995 年间每年仅增长 1%。

Compared with that of last year，the output of 13 main products has increased to a great extent this year。

译文：与去年相比，今年 13 项主要产品的产量都有大幅度增长。

The environmental engineer, on the other hand, must accept the carrying capacity of a stream, an airshed, or a landmass because these can seldom be changed.（见 Unit 4）

译文：另一方面，环境工程师必须掌握河流、大气和土壤的环境容量，因为环境容量很少改变。

（2）比较表达

常用的典型句型有："Noun +be like/the same as/ similar to that one"；"Noun +V +the same quality noun(height/length/price/size/weight) as +noun" "more than" 等。例如：

The cost of one day in an average hospital can run as high as $250.

EIA is applied in more than 100 countries, and by all development banks and most international aid agencies.（见 Unit 4）

（3）倍数和百分数的表达

英语中倍数的增加和减少都可用"increase(decrease) *n* times"表示。还可以用："multiple +as much/many as" 和"increase/decrease +by +number/multiple+"等表示。

The weight of the product is a three times lighter than that of the ordinary.

译文：这个产品的重量是普通产品的 1/3。

Fresh fruit costs twice as much as canned fruit.

译文：新鲜水果比罐头水果贵 1 倍。

The prime cost decreases by 40% (to 60%).

译文：主要成本降低了 40%（到 60%）。

Scientists estimate that for each 1 percent decline in ozone levels, humans will suffer as much as a 2 to 3 percent increase in the incidence of certain skin cancers. (见 Unit 6)

译文：科学家估计臭氧层每减少 1%，人类患皮肤癌的发病率增加 2%～3%。

Exercises

Ⅰ. Point out true or false.

1. Only heavy metals can be monitored by using biological means.

2. Chlorosis of the leaves in plants is caused by sulphur dioxide contamination of the atmosphere.

3. Plants and animals, especially small animals are used to monitor pollution on land.

4. Small mammals exhibit increased body burdens of contamination near to pollution and the extent changes with the source intensity and distance of pollution.

5. Broad leafed plants give a better indication of contamination than narrow leafed ones.

6. Vascular plants are primarily used for monitoring pollution due to their visual symptoms.

7. Small mammals always show consistent body burdens of contamination near pollution sources.

8. Herbage generally exhibits a better response to pollution monitoring compared to plants.

9. The concentrations of pollutants in soil samples always represent their biological availability.

10. Surface contamination on plants represents aerial pollution rather than soil pollution.

Ⅱ. Answer the following questions according to the text.

1. What are the immediate effects of Indoor Air Pollution?

2. What do the long-term effects include?

3. Who are the sensitive proportions of the polulation?

Ⅲ. Translate and fill in the blanks

1. Chlorosis(yellowing) of the leaves in plants is an example which could be due either to soil nutrient deficiency or SO_2 contamination of the atmosphere.

2. 一直到 19 世纪，人们才认识到热是一种能的形式。

3. Significant proportions of the population have a greater sensitivity to pollutants.

4. 像爪子一样，其结构呈环形。

5. 土壤中镉的植物可利用性与土壤中全镉、有机质和速效磷呈显著的正相关关系，而与土壤 pH 值呈显著的负相关关系。

6. A significant change of the average family expenditure in United States ______（出现）since 1995.

7. In 1990, housing ______（占）28% of the average family income, recreation 23%, food and drink 22% and other expenditures 27%. In comparison with 1990, the average family expenditure in 1995 on housing ______（增加）by 7%, on recreation by 12%, but on food and drink ______（减少）by 4%.

8. 总之，研究结果强调了持续监测的重要性，以跟踪污染水平随时间的变化。

9. 对来自不同地点的土壤样品的分析表明，铜元素积累较为显著，尤其是在工业园区附近。

10. On account of this we can find that the average family income in USA ______（增长）from $18000 per year in 1990 to $21000 in 1995.

Ⅳ. Scan the QR code to read the article "Monitoring Ensures Water Protection", and then discuss what professional skills and qualities do you need to become an environmental monitoring worker, and how do advanced technologies in environmental monitoring contribute to the overall effectiveness of environmental regulation?

Unit 6

Air Pollution

【重点与难点】

（1）常用构词法；（2）“see”在科技文献中的应用；（3）“许多”的几种表达方式；（4）汉译英的技巧。

【学习指南】

依据课程重难点，结合课程线上资源及教材的课后习题完成课前预习和课后复习，掌握科技文献互译的典型句型，结合大气污染防治要求与规范，自觉培养生态文明与可持续发展意识。

Text 6

The Ozone Depletion

For four months of every year, Antarctica's McMurdo Research Station lies shrouded in darkness[①]. Then the first rays of light peek out over the horizon. Each day, the sun lingers in the sky just a little longer and the harsh polar winter slowly gives way to spring.

Spring also brings another type of light to the Antarctic, a light that harms instead of nurtures. It is season of new beginnings, the hole in the ozone layer reforms, allowing lethal ultraviolet radiation to stream through Earth's atmosphere.

The hole lasts for only two months, but its timing could not be worse. Just as sunlight awakens activity in dormant plants and animals, it also delivers a dose of harmful ultraviolet radiation. After eight weeks, the hole leaves.

Antarctica is nearby more populated areas, including New Zealand and Australia. This biologically damaging, high-energy radiation can cause skin cancer, injure eyes, harm the immune system, and upset the fragile balance of an entire ecosystem.

Although, two decades ago, most scientists would have scoffed at the notion that industrial chemicals could destroy ozone high up in the atmosphere, researchers now know that chlorine creates the hole by devouring ozone molecules. Years of study on the ground, in aircraft, from satellites have conclusively identified the source of the chlorine: human-made chemicals called chlorofluorocarbons (CFCs) that have been used in spray cans, foam packaging, and refrigeration materials.

Ozone is a relatively simple molecule, consisting of three oxygen atoms bound together. Yet it has dramatically different effects depending upon its location, Near earth's surface, where ozone comes into direct contact with life forms, it primarily displays a destructive side. Because it reacts strongly with other molecules, large concentrations of ozone near the ground prove toxic to living things. At higher altitudes, where 90 percent of our planet's ozone resides, it does a remarkable job of absorbing ultraviolet radiation. In the absence of this gaseous shield in the stratosphere, the harmful radiation has a perfect portal through which to strike earth. Although a combination of weather conditions and CFC chemistry conspire to create the thinnest ozone levels in the sky above the South Pole, CFCs are mainly released at northern latitudes—mostly from Europe, Russia, Japan, and North America—and play a leading role in lowering ozone concentrations around the globe.

Worldwide monitoring has shown that stratospheric ozone has declined for at least two decades, with losses of about 10 percent in the winter and spring and 5 percent in the summer and autumn in such diverse locations as Europe, North America, and Australia. Researchers now find depletion over the North Pole as well, and the problem seems to be getting worse each year. According to a United Nations report, the annual dose of harmful ultraviolet radiation striking the northern hemisphere rose by 5 percent during the past decade.

During the past 40 years, the world has seen an alarming increase in the incidence of malignant skin cancer; the rate today is tenfold higher than in the 1950s. ② Although the entire increase cannot be blamed on ozone loss and increased exposure to ultraviolet radiation, there is evidence of a relationship. Scientists estimate that for each 1 percent decline in ozone levels, humans will suffer as much as a 2 to 3 percent increase in the incidence of certain skin cancers.

CFCs were invented about 65 years ago during a search for a new, nontoxic substance that could serve as a safe refrigerant. One of these new substances, often known by the Dupont trademark Freon, soon replaced ammonia as the standard cooling fluid in home refrigerators. It later became the main coolant in automobile air conditioner.

The 1950s and 1960s saw CFCs used in a variety of other applications: as a propellant in aerosol sprays, in manufacturing plastics, and as a cleanser for electronic components. ③

All this activity doubled the worldwide use of CFCs every six to seven years. By the early 1970s, industry used about a million tons every year.

Yet as recently as the late 1960s, scientists remained unaware that CFCs could affect the atmosphere. Their ignorance was not from lack of interest, but from lack of tools. Detecting the minuscule concentrations of these compounds in the atmosphere would require a new generation of sensitive detectors.

After developing such a detector, the British scientist James Lovelock, in 1970, became the first to detect CFCs in the air. He reported that one of these compounds, CFC-11, had an atmospheric concentration of about 60 parts per trillion. Lovelock found CFC-11 in every air sample that passed over Ireland from the direction of London. That was not surprising, because most major cities, including London, widely used CFCs. However, Lovelock also detected CFC-11 from air samples directly off the North Atlantic, uncontaminated by recent urban pollution.

The scientists showed that CFCs remained undisturbed in the lower atmosphere for decades. Invulnerable to visible sunlight, nearly insoluble in water, and resistant to oxidation, CFCs display an impressive durability in the atmosphere's lower depths. But at altitudes above 18 miles, with 99 height, the harsh, high-energy ultraviolet radiation from the sun impinges directly on the CFC molecules, breaking them apart into chlorine atoms and residual fragments.

In examining these fragments, Rowland and Molina were aided by prior basic research on chemical kinetics—the study of how quickly molecules react with one another and how such reactions take place. Scientists have demonstrated that a simple laboratory experiment would show how rapidly a particular reaction takes place, even if the reaction involves the interaction of a chlorine atom with methane at an altitude of 18 miles and a temperature of –60 degrees Fahrenheit.

After reviewing the pertinent reactions, the two researchers determined that most of the chlorine atoms combine with ozone, the form of oxygen that protects earth from ultraviolet radiation. When chlorine and ozone react, they form the free radical chlorine oxide, which in turn becomes part of a chain reaction. As a result of that chain reaction, a single chlorine atom can remove as many as 100,000 molecules of ozone.

Painstaking research on ozone and the atmosphere over the past 40 years has led to a global ban on CFC production. Since 1987, more than 150 countries have signed an international agreement, the Montreal Protocol, which called for a phased reduction in the release of CFCs so that the yearly amount added to the atmosphere in 1999 would be half that of 1986. Modifications of that treaty called for a complete ban on CFCs that began in

January 1996. Even with this ban in effect, chlorine from CFCs will continue to accumulate in the atmosphere for another decade. It may take until the middle of 21 century for ozone levels in the Antarctic to return to 1970s levels.

New Words and Expressions

Antarctica	*n.* 南极洲，南极大陆
shrouded	*adj.* 覆盖的，遮蔽的
linger	*vi.* 逗留，徘徊，缓慢消失，苟延残喘，奄奄一息
harsh	*adj.* 严厉的，严酷的，刺耳的，现眼的，粗糙的
nurture	*vt.* 养育，培育，滋长，助长；*n.* 教养，培育
ultraviolet	*n.* ；*adj.* 紫外线(的)，紫外辐射(的)
timing	*n.* 时机，时间的选择，时机掌握
dormant	*adj.* 休眠期的，睡着的
scoff	*n.* ；*vi.* 嘲笑
devour	*n.* 吞没，吞噬
chlorine	*n.* 氯，氯气
chlorofluorocarbon	*n.* 氟氯烃
propellant	*n.* 推进剂
aerosol	*n.* 气溶胶，气雾剂
stratosphere	*n.* 同温层，平流层
conspire	*vi.* 共同导致，巧合
malignant	*adj.* 恶性的，致命的
Freon	*n.* 氟利昂
ammonia	*n.* 氨，氨水
minuscule	*adj.* 非常小的，极不重要的
impinge	*vi.* 对……有明显作用，妨碍，侵犯
pertinent	*adj.* 有关的，中肯的，相宜的
Montreal Protocol	蒙特利尔议定书，蒙特利尔公约

Notes

① The problem for four months of every year, Antarctica's McMurdo Research Station lies shrouded in darkness.

麦克默多科考站(McMurdo Research Station)位于美国南极罗斯岛(Ross Island)，是所有南极考察站中规模最大的一个，也是美国其他南极考察站的综合后勤支援基地，有“南极第一城”的美称。

②During the past 40 years, the world has seen an alarming increase in the incidence of malignant skin cancer; the rate today is tenfold higher than in the 1950s.

在过去的 40 年里，全世界恶性皮肤癌的发病率急剧增加，比 20 世纪 50 年代高出 10 倍以上。

③The 1950s and 1960s saw CFCs used in a variety of other applications: as a propellant in aerosol sprays, in manufacturing plastics, and as a cleanser for electronic components.

二十世纪 50 年代和 60 年代，CFCs 广泛地应用作其他产品：如作为气溶胶和塑料制品的推进剂，以及电子元件的清洁剂。

Reading Material 6

Health Impacts of Air Pollution in China

The health effects of air pollution have attracted considerable attention in China. In this review, the status of air pollution in China is briefly presented. The impacts of air pollution on the health of the respiratory system, the circulatory system, the nervous system, the digestive system, the urinary system, pregnancy and life expectancy are highlighted. Additionally, China's actions to control air pollution and their effects are briefly introduced. Finally, the challenges and perspectives of the health effects of air pollution are provided. We believe that this review will provide a promising perspective on the health impacts of air pollution in China, and further elicit more attention from governments and researchers worldwide.

Fresh air is of great significance for human life. However, with the rapid development of modern industrialization, the air quality around us is increasingly affected, especially in developing countries. The WHO has estimated that approximately 2.4 million people die of diseases related to air pollution per year worldwide (Li et al. 2019b). Therefore, air pollution has generated considerable attentions at a global level. At the same time, the Chinese government and related researchers have recently begun to pay close attention to air pollution and its effects. In particular, the impacts of air pollution on human health in China have been explored, and considerable achievements have been made (Zhang et al. 2010; Yu et al. 2013; Lin et al. 2017; Niu et al. 2017; Liu et al. 2018; Yang et al. 2020a). Therefore, a review presenting the health effects of air pollution in China is urgently needed and may be meaningful for the related researchers and governments.

In the remainder of this paper, we first outlined the status of air pollution in China. Then the health impacts of air pollution in China are provided in the 3rd section; we respectively analyzed the impacts of air pollution on the health of the respiratory system, circulatory system, nervous system, digestive system, urinary system, and pregnancy. In the 4th section, China's actions to control air pollution and its effects are briefly introduced. In the conclusion section, we present the challenges and perspectives of the health effects of air pollution.

In China, the main air pollutants present are sulfur dioxide (SO_2), carbon monoxide (CO), nitrogen oxide (NO_x), ozone (O_3), polycyclic aromatic hydrocarbons (PAHs), particulate matter (PM) and so on (Song et al. 2017). Specifically, SO_2 is a type of highly irritating toxic gas, which comes mainly from the combustion of raw coal and other fuels containing sulfur. CO is the product of incomplete combustion of carbon-containing fuels. NO_x is mainly derived from the emissions of motor vehicles and the combustion of nitrogen-containing compounds. Ozone comes from the chemical reactions of CO, NO_x and other compounds in sunlight (Sáenz-de-Miera 2013). PAHs are mainly emitted from coal combustion, traffic and biomass combustion in China (Zheng et al. 2019). And PMs mainly come from direct emissions from combustion, such as power plant and diesel engine exhausts. These air pollutants have been confirmed to be toxic to human health.

Because China is a coal-burning country, the levels of SO_2, NO_x and total suspended particulates (TSP) in the air are higher than those in many other countries. In the period of 1995—2014, the levels of SO_2 and TSP exhibited a downward trend. However, the concentration of NO_x remained stable. And the particulate matter with aerodynamic diameter of 2.5 mm (PM2.5) and O_3 levels further increased even in central and eastern China, which should receive close attention because PM2.5 and O_3 are more harmful to human health than SO_2 and NO_x (Jin et al. 2016; Ruan et al. 2019). These pollutant distributions exhibited spatiotemporal differences. Yang et al. demonstrated that emissions and meteorological variations mainly affected the air quality in western China. They measured six air pollutants (PM2.5, PM10, SO_2, NO_2, CO, and O_3) at 23 sites in western China for 1 year and observed that over highly populated mega-city regions, such as Sichuan and Guanzhong basins, there are high pollutant concentrations, and the Tibetan Plateau exhibited low levels of pollutants. At the same time, these pollutants also had seasonal distribution variations. The concentrations of PM2.5, PM10, SO_2, NO_2 and CO were high in winter and low in summer, whereas the level of O_3 in low altitude regions showed the opposite seasonal trends. An increasing trend was still persistent in O_3 concentrations. Moreover, lighting and stratospheric transmission may have led to the gradual increase in NO_2 levels in the upper atmosphere (Yang et al. 2020b). Zhao's group performed a comprehensive analysis of the spatiotemporal variation of urban air pollution in China based on the data of more than 300 cities from May 2014 to December 2018. They observed that the air pollutants exhibited distinct spatial distribution variations with high levels of NO_2 in north China and east China, high levels of CO in north China and north-west China, high levels of SO_2 in north China and north-east China, high levels of PM2.5 in north China and central China and high levels of PM10 in north-west China. The worst regions for air pollution were the north China plain and in cities in central and western Xinjiang Province. These spatial distributions may be attributed to differences in emissions (Fan et al. 2020a). In addition, Miao et al. analyzed a panel of data for 30 provinces in China and concluded that the regional atmospheric environmental inefficiency was related to SO_2 emissions from industrial soot and NO_x emissions from vehicle exhaust (Miao et al. 2019). Comfortingly, green space is able to reduce air pollution because the leaves of vegetation can absorb gaseous pollutants and induce the deposition of PMs (Kumar et al. 2019). For instance, Wang's group observed that street trees may contribute more to reducing air pollution than grasses in terms of the analysis of the relationship between residential greenness, air pollution and the psychological well-being of urban residents in Guangzhou, China (Wang et al. 2019d).

Notably, Chinese government has taken various measures to control air pollution in the past few years, resulting in great changes in the levels of air pollutants. The levels of PMs, SO_2 and NO_x were reduced. For example, from 2013 to 2016, the annual country-wide mean PM10 and PM2.5 levels significantly dropped by 29% and 42%, respectively, though the concentrations were still higher than the WHO guidelines. Additionally, NO_x and SO_2 decreased by 42% and 50%, respectively. However, O_3 concentrations are still increasing throughout the country, and strict strategies should be taken to control air pollution (Zeng et al. 2019b). In urban areas, the annual average levels of PM2.5, PM10, SO_2, and CO declined from 2015 to 2019. However, the annual mean concentration of O_3 increased, and the rate of increase was nearly 14 times that of the global rate (Fan et al. 2020b). However, the PAHs levels were distributed differently in different places in China. For example, the PAHs concentrations in Beijing and Taiyuan first exhibited an increasing tendency and then a decreasing tendency, whereas a decreasing tendency was observed in the concentrations of particulate PAHs in Nanjing and Guangzhou during 2000—2015 (Yan et al. 2019).

Therefore, the air pollution in China exhibited distinct differences in spatial and temporal distributions, which may be related to emissions, meteorological variations, geographical differences, human activities and government policies.

汉译英技巧

英语句法结构重视“形合”，句中各个组成部分(词、词组、短信)的结合，都用关联词表达其相互关系（如并列、主从、转折、因果、让步等），因而往往组成结构严谨的复合句。汉语的句法结构重视“意合”，句中各个成分借助意义连接起来，很少使用关联词，因此，往往组成比较简单的并列句或分列单句。汉语句子在很多时候都类似于一种“并列结构”，这类句子翻译成英语时，需要转化为符合英语习惯的复合句结构或具有层次结构的单句。翻译技巧可归纳如下：

1. 条件(原因)服从结果

下面的例句没有关系连词，通过分析能找出它隐含的条件和结果关系。

例如：废水处理厂每天排放超过 3780m^3 的水，其中总磷浓度减小到 1.0mg·L^{-1}，分散源负荷减少约 50%就能满足该湖泊水质控制指标。

前句是条件，后句是结果，翻译成：

If concentrations of TP in effluents of wastewater treatment plants with a flow greater than 3780 m^3 per day are reduced to 1.0 mg·L^{-1}, diffuse-source loads would have to be reduced an estimated 50% to meet the target loads for lake water quality improvement.

例如：需要解决的问题日益增多而且复杂，不同学科之间的关系日益密切，于是便建立了庞大的研究队伍。

译文：The increasing magnitude and complex of the problem to be solved and growing connection of different disciplines led to the setting of research teams.

2. 次要服从主要

分析汉语句子结构，确定句子的核心含义，将之翻译为英语的主句部分，其他的次要部分可以短语的形式表达，例如：

据世界卫生组织估计，每年约 2.5 亿人受肠道寄生虫感染，导致 500～1000 万人死亡。

译文：The World Health Organization estimates that 250 million cases of water related diseases such as cholera arise annually, resulting in 5 to 10 million deaths.

分析：“每年约 2.5 亿人受肠道寄生虫感染”为要表达的中心，可由“估计”的宾语子句表达；“导致 500～1000 万人死亡”是围绕中心的补充说明，为此，次要部分采用分词短语补充说明。

3. 方法服从行动

句子表达的含义为通过某种方法而达到目标时，应把“目标”作为主句部分，例如：

本施工期环境评估报告充分利用勘测资料，进行认真的分析，得出了合理的结论。

译文：Making full use of reconnaissance（勘测）data and carrying through analyses, this report of environmental impact assessment for construction term now turns out reasonable conclusions.

分析：“充分利用勘测资料，进行认真的分析”是方法、手段，原句主体是“本施工期环境评估报告得出了合理的结论”。主体译成一个“主语+动词+宾语”单句，而方法和手段部分用短语形式表达。

4. 背景服从判断

当汉语的并列句表达的是背景与判断的关系时，以表达“判断”的部分作为句子的主句部分，例如：

该研究利用计算机模拟模型和 GIS、自动输入参数、输出图表结果，提供了在较大流域范围运用 AGNPS 进行非点源污染控制活动的实例。

译文：Use of computer simulation model and GIS in a large watershed, automatable input process and tabular and spatial outputs, this research offers an example to test the applicability of AGNPS for planning and managing NPS pollution control activities on a regional scale.

分析：把确定为背景的“利用计算机模拟模型和 GIS、自动输入参数、输出图表结果”用名词性结构表达，以突出它们的客观性。

5. 否定服从肯定

科技文献要求言简意赅，否定服从肯定，例如：

坡长λ是水平投影长度，不是平行于土壤坡面的距离。

译文：The slope length λ is the horizontal projection, not the distance parallel to the soil surface.

Exercises

Ⅰ. Point out true or false.

1. Antarctica’s McMurdo Research Station was in darkness in winter.

2. Spring brings not only light to the Antarctic, but also lethal ultraviolet radiation to the dormant plants and animals.

3. Chlorofluorocarbons (CFCs) that have been used in spray cans, foam packaging, and refrigeration materials chemicals could destroy ozone in the troposphere.

4. Ozone in the troposphere can shield ultraviolet radiation, so it can protect the ecosystem balance.

5. The hole in the ozone layer allows beneficial ultraviolet radiation to reach earth's surface.

6. Ozone primarily displays a constructive side near earth's surface, benefiting living things.

7. Chlorine is responsible for the depletion of the ozone layer due to its reaction with ozone molecules.

8. Ozone loss has been linked to an increase in skin cancer, but the relationship is not well-established.

9. CFCs were invented as a safe refrigerant to replace ammonia.

10. Montreal Protocol led to a complete ban on CFCs in the atmosphere by 1996.

Ⅱ. Answer the following questions according to the text.

1. What are the seasonal changes that occur in Antarctica and how do they relate to the formation of the ozone hole?

2. Explain the global efforts to address the issue of ozone depletion, including the Montreal Protocol. What has been achieved through these efforts?

3. Point out the main function of ozone in the atmosphere including the troposphere and stratosphere.

4. What causes the depletion of ozone layer?

5. What challenges still remain?

Ⅲ. Translate the following sentences.

1. 我们相信，通过我们的努力，可以促进贸易，增进友谊。我们乐于与你们商号建立业务关系。(次要服从主要)

2. 环境工程师除了具备数学、物理和工程科学的知识外，还必须具备化学和微生物学知识领域的扎实基础。(次要服从主要)

3. 环境工程师的独特作用在于运用所有可得到的废弃物净化技术，在生物学和技术学之间架起一座桥梁。(方法服从行动)

4. 环境工程和环境保护是两个不同的概念，并非一样。(否定服从肯定)

5. 随着工业革命的到来，人类比过去更能满足自己对空气、水、食物和住所等的需要。(条件服从结果)

6. Although we value the job satisfaction of our employees, improving the productivity

of our company is still of the utmost importance. We will consider the views of our employees to find a balance. (Secondary follows primary)

7. In order to reduce energy consumption, we have promoted the use of green energy within the company, such as solar panels and energy-saving lighting. (Method follows action)

8. Despite the small scale of the project, its impact has caused widespread concern in the industry. (The negative follows the positive)

9. With the increasing awareness of environmental protection, people begin to take more measures to reduce pollution, thus improving air and water quality. (Condition follows result)

10. Even as land resources are under pressure, environmental engineers are exploring innovative soil remediation methods to restore contaminated land to sustainable use. (Background subject to judgment)

Ⅳ. Scan the QR code to read the article "Consistent Efforts on Air Quality Pay off", and then discuss how can individuals and communities actively participate in air pollution control and foster a culture of ecological civilization and sustainable development in order to face more serious challenges on a global scale?

Unit 7 Water Pollution

【重点与难点】

（1）污水处理方面常用的专有名词和缩略词；（2）常用构词法；（3）复习“as”的用法；（4）标题的翻译方法和注意事项。

【学习指南】

依据课程重难点，结合课程线上资源及教材的课后习题完成课前预习和课后复习，掌握标题的写作思路与方法，结合水污染防治要求与规范，自觉培养民族自豪感与使命感。

Text 7

Water Pollution

Comprising over 70% of the earth's surface, water is undoubtedly the most precious natural resource that exists on our planet. Without water, life on the earth would be non-existent: it is essential for everything on our planet to grow and prosper. Although we recognize this fact, we disregard it by polluting our rivers, lakes, and oceans. Subsequently, we are slowly but surely harming our planet to the point where organisms are dying at a very alarming rate. In addition to innocent organisms dying off, our drinking water has become greatly affected, as is our ability to use water for recreation purposes. In order to combat water pollution, we must understand the problems and become part of the solution.

Many causes of pollution, including sewage and fertilizer, contain nutrients such as nitrates and phosphates. In excess levels, nutrients over stimulate the growth of aquatic plants and algae. Excessive growth of these types of organisms consequently clogs our waterway, uses up dissolved oxygen as they decompose, and blocks light to deeper waters. This, in turn, proves very harmful to aquatic organisms as it affects the respiration ability of fish and other invertebrates that reside in water.

Pollution is also caused when silt and other suspended solids, such as soil, wash off plowed fields, construction and logging sites, urban areas, and eroded riverbanks when it rains. Eutrophication is an aging process that slowly fills in the water body with sediment and organic matter. When these sediments enter various bodies of water, fish respiration becomes impaired, plant productivity and water depth become reduced, and aquatic organisms become suffocated. Pollution in the form of organic material enters water in many different forms as sewage, as leaves and grass clippings, or as runoff from livestock feedlots and pastures. When natural bacteria and protozoan in the water break down this organic material, they begin to use up the oxygen dissolved in the water. Many types of fish and bottom-dwelling animals cannot survive when levels of dissolved oxygen drop below two to five parts per million. When this occurs, it kills aquatic organisms in large numbers that leads to disruptions in the food chain.

The major sources of water pollution can be classified as municipal, industrial, and agricultural. Municipal water pollution consists of wastewater from homes and commercial establishments. For many years, the main goal of treating municipal wastewater was simply to reduce its content of suspended solids, oxygen-demanding materials, dissolved inorganic compounds, and harmful bacteria. In recent years, however, more stress has been placed on improving means of disposal of the solid residues from the municipal treatment processes. The basic methods of treating municipal wastewater fall into three stages: primary treatment, including grit removal, screening, grinding, and sedimentation; secondary treatment, which entails oxidation of dissolved organic matter by means of using biologically active sludge, which is then filtered off; and tertiary treatment, in which advanced biological methods of nitrogen removal and chemical and physical methods such as granular filtration and activated carbon absorption are employed. ① The handling and disposal of solid residues can account for 25 to 50 percent of the capital and operational costs of a treatment plant.

The characteristics of industrial wastewater can differ considerably both within and among industries. The impact of industrial discharges depends not only on their collective characteristics, such as biochemical oxygen demand and the amount of suspended solids, but also on their content of specific inorganic and organic substances. Three options are available in controlling industrial wastewater: (1) control can take place at the point of generation in the plant, (2) wastewater can be pretreated for discharge to municipal treatment sources, (3) wastewater can be treated completely at the plant and either reused or discharged directly into receiving waters.

Agriculture, including commercial livestock and poultry farming, is the source of many organic and inorganic substances in surface water and groundwater. These contaminants include both sediment from erosion cropland and compounds of phosphorus and nitrogen that partly originate in animal wastes and commercial fertilizers. Animal wastes are high in oxygen demanding material, nitrogen and phosphorus, and they often harbor pathogenic organisms. Wastes from commercial feeders are contained and disposed of on land; their main threat to natural waters, therefore, is from runoff and leaching. Control may involve settling basins for liquids, limited biological treatment in aerobic lagoons, and a variety of other methods.

New Words and Expressions

comprise	*vt.* 包含，构成，由……组成
innocent	*adj.* 无罪的，无辜的
combat	*vt.* 跟……战斗，反对；*n.* 格斗，战斗，争论
aquatic	*adj.* 水生的，水产的，水栖的
algae	*n.* (pl.) 水藻，海藻
clog	*vt.* 妨碍，填满（管子、道路）；*n.* 障碍，障碍物
respiration	*n.* 呼吸（作用）
invertebrate	*adj.* 无脊椎的，无脊椎动物的
silt	*n.* 淤泥，（河边等的）淤泥沉积处
erode	*vi.* 被腐蚀，被侵蚀；*vt.* 腐蚀，侵蚀
eutrophication	*n.* 富营养化
impaired	*adj.* 受损伤的
livestock	*n.* 家畜
feedlot	*n.* 饲养场
suffocate	*vt.* 使窒息，闷死；*vi.* 窒息，呼吸困难
pasture	*n.* 牧草场；*vt.* 放牧，吃草
protozoan	*n.* 原生动物
filter	*n.* 滤器，滤池，滤纸
granular	*adj.* 粒状的，含颗粒的
grit	*n.* 砂粒，硬渣
entail	*vt.* 使成为必要，需要
nitrate	*n.* 硝酸盐
nutrient	*adj.* 营养的，滋养的；*n.* 营养品，营养物

option *n.* 自由选择，选择权
pathogen *n.* 病原体，病原菌
plow *n.*; *vt.*; *vi.* (=plough) 犁，耕地
phosphate *n.* 磷酸盐
prosper *vi.* 繁荣，昌盛，成功

Notes

① The basic methods of treating municipal wastewater fall into three stages: primary treatment, including grit removal, screening, grinding, and sedimentation; secondary treatment, which entails oxidation of dissolved organic matter by means of using biologically active sludge, which is then filtered off; and tertiary treatment, in which advanced biological methods of nitrogen removal and chemical and physical methods such as granular filtration and activated carbon absorption are employed.

市政废水的处理方法可分为三级处理:一级处理，包括沉砂、筛分、破碎、沉淀；二级处理，主要是利用生物活性污泥对溶解在水中的有机物进行氧化，然后通过过滤去除剩余污泥；三级处理，采用了高级生物方法除氮，也采用了一些如粒状介质过滤、活性炭吸附等的物理化学方法。

Reading Material 7

Effects of Water Pollution on Human Health and Disease Heterogeneity: A Review

Water is an essential resource for human survival. According to the 2021 World Water Development Report released by UNESCO, the global use of freshwater has increased six-fold in the past 100 years and has been growing by about 1% per year since the 1980s. With the increase of water consumption, water quality is facing severe challenges. Industrialization, agricultural production, and urban life have resulted in the degradation and pollution of the environment, adversely affecting the water bodies (rivers and oceans) necessary for life, ultimately affecting human health and sustainable social development (Xu et al., 2022a). Globally, an estimated 80% of industrial and municipal wastewater is discharged into the environment without any prior treatment, with adverse effects on human health and ecosystems. This proportion is higher in the least developed countries, where sanitation and wastewater treatment facilities are severely lacking.

Water pollution are mainly concentrated in industrialization, agricultural activities, natural factors, and insufficient water supply and sewage treatment facilities. First, industry is the main cause of water pollution, these industries include distillery industry, tannery industry, pulp and paper industry, textile industry, food industry, iron and steel industry, nuclear industry and so on. Various toxic chemicals, organic and inorganic substances, toxic solvents and volatile organic chemicals may be released in industrial production. If these wastes are released into aquatic ecosystems without adequate treatment, they will cause water pollution (Chowdhary et al., 2020). Arsenic, cadmium, and chromium are vital pollutants discharged in wastewater, and the industrial sector is a significant contributor to harmful pollutants (Chen et al., 2019). With the acceleration of urbanization, wastewater from industrial production has gradually increased (Wu et al., 2020). In addition, water pollution caused by industrialization is also greatly affected by foreign direct investment. Industrial water pollution in less developed countries is positively correlated with foreign direct investment (Jorgenson, 2009). Second, water pollution is closely related to agriculture. Pesticides, nitrogen fertilizers and organic farm wastes from agriculture are significant causes of water pollution (RCEP, 1979). Agricultural activities will contaminate the water with nitrates, phosphorus, pesticides, soil sediments, salts and pathogens (Parris, 2011). Furthermore, agriculture has severely damaged all freshwater systems in their pristine state (Moss, 2008). Untreated or partially treated wastewater is widely used for irrigation in water-scarce regions of developing countries, and the presence of pollutants in sewage poses risks to the environment and health. The imbalance in the quantity and quality of surface water resources has led to the long-term use of wastewater irrigation in some areas in developing countries to meet the water demand of agricultural production, resulting in serious agricultural land and food pollution, pesticide residues and heavy metal pollution threatening food safety and human health (Lu et al., 2015). Pesticides have an adverse impact on health through drinking water. Comparing pesticide use with health life Expectancy Longitudinal Survey data, it was found that a 10% increase in pesticide use resulted in a 1% increase in the medical disability index over 65 years of age (Lai, 2017). The case of the Musi River in India shows a higher incidence of morbidity in wastewater irrigated villages than normal-water households. Third, water pollution is related to natural factors. Taking Child Loess Plateau as an example, the concentration of trace elements in water quality is higher than the average world level, and trace elements come from natural weathering and manufactured causes. Poor river water quality is associated with high sodium and salinity hazards (Xiao et al., 2019). The most typical water pollution in the middle part of the loess Plateau is hexavalent chromium pollution, which is caused by the natural environment and human activities. Loess and mudstone are the main sources, and groundwater with high concentrations of hexavalent chromium is also an important factor in surface water pollution (He et al., 2020). Finally, water supply and sewage treatment facilities are also important factors affecting drinking water quality, especially in developing countries. In parallel with China rapid economic growth, industrialization and urbanization, underinvestment in basic water supply and treatment facilities has led to water pollution, increased incidence of infectious and parasitic diseases, and increased exposure to industrial chemicals, heavy metals and algal toxins (Wu et al., 1999). An econometric model predicts the impact of water purification equipment on water quality and therefore human health. When the proportion of household water treated with water purification equipment is reduced from 100% to 90%, the expected health benefits are reduced by up to 96%. When the risk of pretreatment water quality is high, the decline is even more significant (Brown and Clasen, 2012).

To sum up, water pollution results from both human and natural factors. Various human activities will directly affect water quality, including urbanization, population growth, industrial production, climate change, other factors (Halder and Islam, 2015) and religious activities (Dwivedi et al., 2018). Improper disposal of solid waste, sand, and gravel is also one reason for decreasing water quality (Ustaoğlua et al., 2020).

Unsafe water has severe implications for human health. According to UNESCO 2021 World Water Development Report, about 829,000 people die each year from diarrhea caused by unsafe drinking water, sanitation, and hand hygiene, including nearly 300,000 children under the age of five, representing 5.3 percent of all deaths in this age group. Data from Palestine suggest that people who drink municipal water directly are more likely to suffer from diseases such as diarrhea than those who use desalinated and household-filtered drinking water (Yassin et al., 2006). In a comparative study of tap water, purified water, and bottled water, tap water was an essential source of gastrointestinal disease (Payment et al., 1997). Lack of water and sanitation services also increases the incidence of diseases such as cholera, trachoma, schistosomiasis, and helminthiasis. Data from studies in developing countries show a clear relationship between cholera and contaminated water, and household water treatment and storage can reduce cholera (Gundry et al., 2004). In addition to disease, unsafe drinking water, and poor environmental hygiene can lead to gastrointestinal illness, inhibiting nutrient absorption and malnutrition. These effects are especially pronounced for children.

写作技巧

标题的写作

科技论文标题的特点是逻辑严谨，结构简洁，主题突出，使读者一看便可知道论文涉及的范围，窥见文章梗概，起到“一叶知秋”的作用。在文摘、索引和题录等二次文献中，一般只列出论文的题目、摘要和出处。例如国际权威的 Index to Scientific and Technical Proceedings（简称 ISTP）、Engineering Index（简称 EI）和 Chemical Abstract（简称 CA）均是如此。

标题一般不是句子，没有完整的主谓结构，只在少数情况（评述性、综述性和驳斥性）下可用疑问句作标题名。标题的译写应当注意以下问题：

1. 抓住中心词

科技论文的标题一般首先提出中心词，中心词以名词为主，然后附加一些修饰语，多为后置定语。而汉语论文标题常用偏正词组，中心词在最后面。例如：

中国经济学的过去与未来

The Past and Future of Chinese Economics

“注意力”的经济学描述

An Economic Approach to the Concept of Attention

2. 抓住纽带词

英、汉两种语言在论文标题中各有自己的纽带词，通过这些纽带词把标题的几个部分连接起来。在汉语中常用的纽带词有“的”“对”“在”“与”“和”等；在英语中常用的纽带词有介词、分词、形容词、动词不定式等。例如：

我国工业产品的生态特征(表示所属关系)

Ecological Characteristics of Industrial Product in China

腐植酸钠对植物吸收磷的影响(表示行为主体)

Effect of Sodium Humate on Uptake of Phosphorus by Plants

3. 标题力求简洁

科学论文标题要力求简洁，同时还要能高度概括全文内容。标题长度一般不超过 12 个词或 100 个书写符号（包括间隔）。按照国际标准，标题以 8 个词为佳。因此标题目应尽量删去多余的词。求得标题简洁的方法有以下几种：

- **删去一些陈旧的标题习语**。如“Some Thoughts on”“A Few Observations on”“Study of”，这些词的省略符合英语标题的潮流，可使标题更加简洁。
- **表示谦逊和科学上留有余地的词语不需译出**。如：研究（a study on, a review of, on）、初探(preliminary analysis on)、探讨、商榷（discussion on）、观察（observation）、

报告（a report of）、分析（analysis）、经验和体会（experience in, personal understanding）等。例如：

自然资源与环境的研究

Natural Resources and Environment

试论中国的人口控制问题

Population Control in China

- **冠词的简化**。科技论文标题中的冠词在早年用得较多，近些年有简化的趋势，凡可用可不用的冠词均可不用。例如：

地下水质对小麦产量和品质的影响

Effect of Groundwater Quality on Wheat Yield and Quality

从废母液中回收皂素的工艺

Techniques for Retrieving Saponin from Waste Mother Liquor

其中两处的冠词 the 均可不用。

4. 注意标题的大小写问题

标题中大小写的使用方法，不同国家，不同刊物略有不同。通常情况是除冠词、介词和连词以外，其余实词的第一个字母都要大写，介词和连词四个字母以上者也可大写，七个字母以上者最好大写。标题每一行的第一个词一律大写，最后一个词是介词时，大小写均可。

此外，国内外也有不少学术刊物采取标题全部用大写字母。

Exercises

Ⅰ. Point out true or false.

1. Air, water and soil are the most precious natural resource on our planet.

2. The controlling of phosphorus has been the key factors leading to the eutrophication of water bodies.

3. In order to meet the effluent standards, municipal wastewater should be treated by secondary treatment.

4. When it is unfit for drinking, water is considered polluted.

5. Eutrophication is a process that clears water bodies of sediment and organic matter.

6. Organic material entering water through sewage has no impact on dissolved oxygen levels.

7. Industrial wastewater characteristics are generally uniform across different industries.

8. The main goal of treating municipal wastewater is to reduce its content of solid residues.

9. Tertiary treatment of wastewater involves only physical methods like filtration.

10. Agriculture is not a significant source of inorganic contaminants in water.

Ⅱ. Answer the following questions according to the text.

1. What are the processes included in primary, secondary and tertiary treatment?

2. How many types can the sources of water pollution be divided into?

3. What are some consequences of excessive growth of aquatic plants and algae due to nutrient pollution?

4. How does pollution in the form of organic material impact dissolved oxygen levels in water?

5. How does industrial wastewater control vary among different options available to industries?

Ⅲ. Translate the following titles.

1. 低浓度 SO_2 烟气的综合治理
2. 防除杂草的问题解决了吗？
3. 正确认识奖金的本质，充分发挥奖金的作用
4. 农业发展的近期与长远设想
5. 科学在社会发展中的重要作用
6. 国外进口对本国工业化的影响
7. 基于物联网监测的城市环境空气质量评估
8. 可再生能源整合在可持续城市发展中的优化应用
9. 塑料废物管理的创新策略：循环经济方法的案例研究
10. 遥感在热带雨林砍伐监测中的应用

Ⅳ. Modify the following titles.

1. Preliminary Discussion about the Chromosome Variation of the pollen mother cells of the pollen-Derived plants in the wheat (*Triticum aestivum L.*)

2. A Discussion about the Relationship Between Pure Science and Applied Science

3. A study of factors influencing the bactericidal efficiency of chlorine

4. The water quality monitoring shows Poyang Lake have been polluted

5. We have proven that geosynthetic cut-off walls are a technology that has the potential to make great contributions in the remediation process of contaminated waste sites

6. New developments in strategies for immobilizing and stabilizing hazardous waste

have emerged and have a place in our future for our protection

Ⅴ. Scan the QR codes and combine them with "Resolve Effects at Protecting China's Mother Rivers", and discuss how can the principles of water pollution prevention and ecosystem protection be futher intergrated into the education system and public awareness to enhance citizens'national pride and responsibility for environmental management?

Unit 8

Solid Waste

【重点与难点】

（1）固体废弃物的相关词汇；（2）常用构词法的复习；（3）几种强调句的用法的复习；（4）摘要构成与写作。

【学习指南】

依据课程重难点，结合课程线上资源及教材的课后习题完成课前预习和课后复习，掌握摘要的写作思路与方法，结合固体废物，特别是危险废物处理处置的典型案例，自觉培养科研报国、勇于创新的精神。

Text 8

Solid Waste

Since the beginning, humankind has been generating waste, be it the bones and other parts of animals they slaughter for their food or the wood they cut to make their carts. With the progress of civilization, the waste generated became of a more complex nature. At the end of the 19th century, the industrial revolution saw the rise of the world of consumers. Not only did the air get more and more polluted but the earth itself became more polluted with the generation of nonbiodegradable solid waste.

Municipal solid waste consists of household waste, construction and demolition debris, sanitation residue and waste from streets. This garbage is generated mainly from residential and commercial complexes. With rising urbanization and change in lifestyle and food habits, the amount of municipal solid waste has been increasing rapidly and its composition has been changing. In 1947, cities and towns in India generated an estimated 6 million tons of solid waste, in 1997, there was about 48 million tons. More than 25% of the municipal solid waste is not collected at all; 70% of the Indian cities lack adequate capacity to transport it and there are no sanitary landfills to dispose of the waste. The existing landfills are neither well equipped nor well managed and are not lined properly to protect against contamination of soil and groundwater.

Over the last few years, the consumer market has grown rapidly leading to products being packed in cans, aluminum foils, plastics, and other such nonbiodegradable items that cause incalculable harm to the environment. In India, some municipal areas have banned the use of plastics and they seem to have achieved success. For example, today one will not see a single piece of plastic in the entire district of Ladakh where the local authorities imposed a ban on plastics in 1998. Other states should follow the example of this region and ban the use of items that cause harm to the environment. One positive note is that in many large cities, shops have begun packing items in reusable or biodegradable waste, which will considerably lessen the burden of solid waste that each city has to tackle.

There are different categories of waste generated, each take their own time to degenerate (as illustrated in Table 8-1).

Industrial and hospital waste are considered hazardous as they may contain toxic substances. Certain types of household waste are also hazardous. Hazardous wastes could be highly toxic to humans, animals, and plants. And they are corrosive, highly inflammable, or explosive, and react when exposed to certain things e. g. gases. India generates around 7 million tons of hazardous wastes every year, most of which is concentrated in four states: Andhra Pradesh, Bihar, Uttar Pradesh, and Tamil Nadu. Household wastes that can be categorized as hazardous waste include old batteries, shoe polish, paint tins, old medicines, and medicine bottles.

Table 8-1 The Type of Litter and the Approximate Time of its Degeneration

No.	Type of litter	Approximate time it takes to degenerate
1	Organic waste (vegetable, fruit peels, leftover foodstuff, ect)	A week or two
2	Paper	10-30 days
3	Cotton cloth	2-5 months
4	Wood	10-15 years
5	Woolen items	1 year
6	Metals	100-500 years
7	Plastics	Undetermined
8	Glass	Undetermined

Hospital waste contaminated by chemicals used in hospitals is considered hazardous. These chemicals include formaldehyde and phenols, which are used as disinfectants and mercury, which is used in thermometers or equipments that measure blood pressure. Most hospitals in India do not have proper disposal facilities for these hazardous wastes.

In the industrial sector, the major generators of hazardous waste are the metal, chemical, paper, pesticide, dye, refining, and rubber goods industries. Direct exposure to chemicals in hazardous waste such as mercury and cyanide can be fatal.

Hospital waste is generated during the diagnosis, treatment, or immunization of human beings or animals or in research activities in these fields or in the production or testing of biological. It may include wastes like sharps, soiled waste, disposables, anatomical waste, cultures, discarded medicines, chemical wastes, etc. These are in the form of disposable syringes, swabs, bandages, body fluids, human excreta, etc. This waste is highly infectious and can be a serious threat to human health if not managed in a scientific and discriminate manner. It has been roughly estimated that of the 4 kg of waste generated in a hospital at least 1 kg would be infected.

New Words and Expressions

slaughter	*n*. 屠宰，（运动）大败；*vt*. 杀戮，屠宰
nonbiodegradable	*adj*. 不可生物降解的
demolition	*n*. 毁坏，破坏，拆毁
sanitation	*n*. 卫生，卫生设施
landfill	*n*. 垃圾填埋场
complex	*n*. 混合物，集合体
lessen	*vt*. 减轻，减少，轻视；*vi*. 变小，缩小，减少
tackle	*vt*. 解决，应付
inflammable	*adj*. 易燃的，易怒的
polish	*n*. 鞋油，上光剂
formaldehyde	*n*. 甲醛
phenol	*n*. 苯酚，石碳酸
disinfectant	*n*. 消毒剂
cyanide	*n*. 氰化物
immunization	*n*. 免疫作用，免疫
anatomical	*adj*. 解剖的，解剖学
syringe	*n*. 注射器，灌肠器，洒水器；*vt*. 注射，灌肠
swab	*n*. 药签，拖把
excreta	*n*. 排泄物

Reading Material 8

Municipal Solid Waste Management: A Review of Waste to Energy (WtE) Approaches

The world is experiencing a rapidly growing population and rising public living standard, which leads to increases in the generation of municipal solid waste (MSW) and consumption of energy and goods. These activities result in changes in land use, deforestation, intensified agricultural practices, industrialization, and energy use from fossil fuels. All of these practices along with the MSW generation lead to ever-growing concentrations of greenhouse gases (GHG) in the environment and higher risk to public health by unscientific disposal (Palacio 2019). MSW can be defined as non-hazardous, biodegradable / non-biodegradable, carbonaceous/non-carbonaceous, and reusable or unusable solid waste that is generated from households, offices, trade, garden, yard, and street (Ngusale et al. 2017; Abbasi 2018). MSW in the developing world is facing significant management and disposal problems. MSW management problems are more prominent in the middle-and low-income countries due to the fast growing population and urbanization (Alam and Ahmade 2013; Damtew and Desta 2015).

Previous work correlates the increase in solid waste generation rate with the increase in population, technological development, and changes in the public lifestyle (Ali 2009; Monavari et al. 2012; Leone et al. 2013). The solid waste management problems are more acute in developing countries than the developed countries (Zerbock 2003; World Bank 2012). Currently, the global population is reaching around 7.1 billion, and it generates around 1300 million tons of solid waste per year. For instance, the MSW is estimated to be around 1.2 kg per capita per day, which is mostly generated from the urban parts of the major cities. The waste amount is expected to reach up to 2200 million tons per year by the year 2025 (World Bank 2012). Hence, proper characterization and management of MSW are the need of the hour. Improper and inefficient solid waste management leads to GHGs emission, odors problems, and high risk to public health. It has been predicted that emission of GHGs from waste management in developing countries will increase exponentially (Friedrich and Trois 2011). Further, researchers (Memon 2010; Rotich et al. 2006) investigated that only 20% to 50% of municipalities in developing countries provide funds for solid waste management, which covers less than 50% of the total population. Therefore, solid waste management has become a hot issue due to environmental risks and challenges, which further has prompted researchers to consider MSW as a resource to be managed in a sustainable manner. Therefore, around the world, to ensure sustainable MSW management, options such as mainly recycling, landfill with gas recovery, thermal, and biochemical conversion methods are being considered (Shekdar 2009). Recently, waste treatment processes generating energy from MSW in the form of heat, electricity, or transport fuels have received special attention and considered as waste to energy (WtE) options. WtE technology includes mainly pyrolysis, gasification, combustion, anaerobic digestion, biohydrogen production, and landfill gas recovery. Among these, thermochemical conversion, anaerobic digestion, and landfilling with energy recovery options are based upon the theme of the energy recovery option in the MSW management hierarchy (Lu et al. 2006).

Previous estimation related to WtE showed that by 2025, 2.3 billion tons of solid waste will be generated annually, which will have huge potential effects on power generation (Islam 2016). Thus, the waste to energy approach is a promising energy alternative option for the future because of having potential to fulfill 10% of total global electricity demand (World Energy Council 2013). Another report suggests the utilization of WtE technology for the treatment of approximately 261 million tons per year of MSW by 2022, which has potential to generate around 283 terawatt (TWh) hours of heat and electricity (World Energy Council 2013). The two most common strategies of WtE techniques are solid waste incineration and landfill gas recovery (LFG) systems (Tan et al. 2014), but the most suitable option from a financial perspective in the future energy system is mixed MSW incineration (Marie and Peter 2011). Some countries such as Denmark and Sweden have already had well-established energy generation systems based on incineration for more than a century. The data showed that waste incineration system produce 4.8% of the total electricity consumption and 13.7% of the total domestic heat consumption in Denmark (Kleis and Dalagar 2007). WtE technologies have received much attention due to significant waste volume reduction along with the renewable energy production to meet the present as well as the future energy demands. Therefore, the authors have planned a systematic and in depth study for solid waste generation, composition and sustainable MSW management using biochemical and thermal energy conversions approaches. Further, the present study can be helpful in the conservation and preservation of natural resources, environment, and public health, which further contribute to the general goals of sustainability.

摘要的写作

为了便于交流，许多以中文发表的科技论文、研究生学位论文、本科生学位论文都要求有长短不同的英文摘要（abstract）。摘要是把原文缩减后形成的脱离原文而独立存在的短文，通常位于标题和正文之间，摘要内容包括论文的“目的（purposes）”，主要的研究“过程（procedures）”及所采用的“方法（methods）”，由此得到的主要“结果（results）”和重要“结论（conclusions）”。

1. 摘要写作注意事项

- 摘要应忠实于正文，不能包含正文中没有的信息资料。
- 要使用正规英语、标准术语，避免使用缩写字。或者说，对第一次出现的缩写字或符号等必须加以注明。注明方式有两种：一种是先缩后注，如 GIS (Geographic Information System), LAN (Local Area Network)；一种是先注后缩，如 Agricultural Non-Pollution Model (AGNPS), Geographic Resource Analysis Support System (GRASS)。
- 行文不用第一人称。如出现“我们”“作者”作为主语，一是会削弱表述的客观性，二是会使英文读者搞不清论文主题到底是作者在做一项科研还是在介绍一个科研成果。此种论文一般不会被编辑刊用。
- 在说明研究目的、叙述研究内容、描述结果、得出结论、提出建议或讨论时采用一般现在时；在叙述发现、研究过程时采用一般过去时；在把已完成的研究与现在结论联系起来时采用现在完成时。

2. 摘要的编写方法

摘要通常按主题句、扩展句和结尾句的顺序构成，这是摘要写作最常用的方法。主题句之后通常有 3～5 个简短的句子，展开摘要的内容，进一步阐述主题句的内容，概括出文章的主要事实和论点，介绍研究的内容、方法和结果。多数用一般过去时被动语态，少数用现在完成时或一般现在时主动语态。当强调某些动作在过去某特定时间之前完成，用过去完成时。摘要的结尾，通常有结束句，对全文作出结论或补充交代，指出其适用范围和有效性，或评价本研究的重要意义和实用价值，提出建议和措施。有的摘要对论文内容列举，直到结束不再对文章内容进行归纳，也就没有结尾句。

主题句阐明论文的主题。常见的主题句有：

A new approach is proposed to …, A new method was developed for …, This paper analysis …, This paper discusses …和 a certain topic …is (are) introduced (proposed, presented, described, studied, conducted)等。

主题句的例子如下：

An Agricultural Non-Point Source Pollution Model (AGNPS), the Geographic Resource Analysis Support System (GRASS) (U.S. Army Corps of Engineers, 1987), and a hydrologic modeling tool box (GRASS WATERWORKS, being developed at the Michigan State University, Center for Remote Sensing）were integrated to evaluate the impact of agricultural runoff on water quality in the Cass River, a sub-watershed of Saginaw Bay.

本研究集成农业非点源污染模型(AGNPS)、地理资源分析支持系统（GRASS）(美国陆军工兵部队，1987）和水文模拟工具箱（GRASS WATERWORKS，密歇根州立大学遥感中心开发），用以评估萨吉诺湾卡斯河子流域农业径流对水质的影响。

展开句的例子如下：

Each year from 1990-1994, three groups of six beef steers, 9 months old, were stocked at 2.6 ha. and managed under one of the following treatments: (A)pasture only; (B)winter feeding; (C)autumn deferment.

有的论文主题句本身就具有展开句的性质，直接引出文章的主要内容。例如：

This paper begins with the discussion on the recent situation of electroplating production and its environmental problems. Next, its developing trend at home and abroad are described. Then it introduces the three primary principles: crude material reduced, recycle and resource to resolve the pollution problems and realize the sustained development in electroplating industry.

本文阐述了电镀工业生产现状及存在的环境问题，介绍了电镀工业国内外技术发展趋势，论述了电镀循环经济三大原则，即“减量化、再利用、资源化”，能从根本上解决电镀生产污染问题，实现电镀工业的可持续发展。

一般性结论用一般现在时，特定条件下的结论用一般过去时，设想和建议用一般现在时，例如：

Consequently, there was no relationship between semi-dwarfing genes and amounts and pattern of water uptake in the dryland winter wheat cultivars investigated.

3. 关键词

一篇规范、严谨的科技论文少不了“关键词”一项。关键词通常用“Keywords”“Key Words”或“Key words”表示。关键词之间可用逗号或分号隔开，或采用大间隔的方法而不用标点符号分隔，最末一个关键词后面一般不加任何标点符号。

科技论文的关键词常作为“keyword index”列入情报文献的检索系统，有利于全文的检索和流通。关键词一般为4～6个，最少2个，最多不超过8个。一般用名词或名词词组形式，不用动词形式。比如“分析”用“analysis”而不用“analyse”。关键词可以从论文的标题和摘要中选择，因为这是论文中最关键的部分。但选词必须规范，应符合ISO（国际标准化组织）规定的国际标准，以利于全文的检索和流通。

Exercises

Ⅰ. Point out true or false.

1. At the end of the 19th century, the increase in population and urbanization was

responsible for responsible for the increase in solid waste.

2. Municipal solid waste consists of food wastes, construction and demolition debris, rubbish and sanitation residue.

3. Today nobody will see a single piece of plastic in big city of India, for the local authorities imposed a ban on plastics in 1998.

4. Hospital waste is highly infectious, so it should be treated and managed in a scientific and discriminate manner.

5. The rise of consumerism during the industrial revolution led to a decrease in waste generation.

6. Municipal solid waste includes waste from industries and factories.

7. In India, more than 70% of cities have adequate capacity to transport their municipal solid waste.

8. Hazardous wastes are never toxic to humans, animals, or plants.

9. Mercury is commonly used in hospitals for waste disposal.

10. Plastics and glass have an undetermined time of degeneration.

Ⅱ. Answer the following questions according to the text.

1. How many types of solid wastes are there?

2. What is solid waste?

3. How did the amount of municipal solid waste change from 1947 to 1997 in India?

4. What measures were taken in Ladakh to eliminate plastic waste?

5. Why is hospital waste considered highly infectious and dangerous to human health?

Ⅲ. Translate the following abstract.

超声光催化降解苯胺及其衍生物研究

顾浩飞[1]，安太成[1,2 3]，文晟[2]，陈卫国[1]，朱锡海[1]，傅家谟[2]，盛国英[2]

(1. 中山大学化学与化学工程学院，广州 510275；

2. 中国科学院广州地球化学研究所有机地球化学国家重点实验室，广东；

省环境资源利用与保护重点实验室，广州 510640)

摘要：以苯胺及其衍生物为研究对象，探讨了不同有机化合物结构对超声光催化降解的影响，进一步开拓了基于超声波与 TiO_2 光催化联合催化降解有机废水的新型深度氧化处理新技术——超声光催化技术。实验结果表明，尽管超声光催化反应对苯胺及其衍生物的降解协同效应并不是很显著，但是超声光催化反应对苯胺及其衍生物却具有较好的降解率，而且不同有机化合物结构对超声光催化反应有着较大的影响。

关键词：超声光催化；光催化；超声波；分子结构；结构活性

Ⅳ. Modify the following title and abstract.

南昌地区不同企业周边重金属分布及影响规律

本文结合 GIS 技术和主成分分析方法，综合评价南昌地区钢铁、化工、电镀和塑料 4 种类型 6 个不同企业周边红壤-蔬菜系统的环境质量。结果表明，四种元素的污染程度依次为：Cd>Cu>Zn>>Pb，其中 Pb 含量未超过土壤环境质量二级标准。重金属 Cd、Zn 和 Cu 具有良好的同源性，其分布均是呈辐射状，从南到北逐渐降低，东南方向的电镀厂和钢铁厂 2 周边区域为最高浓度区。企业周边蔬菜地中白菜 (Brassica chinensis) 未受 Pb 污染，Cu、Zn 和 Cd 已在白菜体内达到了一定程度的积累。白菜中重金属与土壤中对应重金属含量显著相关，说明蔬菜中重金属主要来源于表层土壤中所积累的重金属。

Distribution and Influence of Heavy Metal around Different Enterprises in Nanchang Area

Combined with GIS (Geographic Information System) technology and PCA (principal component analysis) method, the environment quality of red soil-vegetable system around 6 different enterprises including iron and steel, chemical industry, electroplating and plastic enterprises in Nanchang area was assessed. The results indicated that among each heavy metal, degree of pollution was represented as a trend of Cd>Cu>Zn>>Pb, and soil Pb content was below the second grade of Standards for Soil Environmental Quality of China (GB15618-1995) in all samples. There was significant homology among Cd、Zn and Cu. On the spatial dynamics of heavy metal transportation in soil, heavy metals content in surface layer was down from south to north. The highest content area was southeast because electroplating and iron and steel plant 2 were there. The contents of Cd、Zn and Cu in pakchoi (Brassica chinensis) were above the Hygienic Standards for food security of P.R. China, but the content of Pb was reverse. There were significant relationship of the heavy metal content between soil and pakchoi, for heavy metal in top soil was the main source.

Ⅴ. Scan the QR code to read the article "Waste Sorting Work Gets a Major Push", and then discuss faced with the challenge of managing inadequate facilities, how can scientific research and innovative technologies contribute to improving the disposal and treatment of hazardous waste?

Unit 9

Noise

【重点与难点】

（1）长句的翻译方法；（2）噪声控制方面的常用词汇；（3）摘要构成与写作。

【学习指南】

依据课程重难点，结合课程线上资源及教材的课后习题完成课前预习和课后复习，掌握摘要的写作思路与方法，结合噪声的处理处置规范，自觉培养人与自然和谐共生的意识。

Text 9

Noise

The word "noise" carries the meaning of unwanted sound. This interpretation implies a value judgment of the sound, which in turn generally implies the response of a human being to a noise environment. In our study, we will be concerned with noise environments which have an adverse effect on human beings.

We are all aware that a very intense noise can cause permanent hearing loss, but perhaps we have less appreciation for the fact that the same type of hearing damage can accrue after exposure to more moderate noise environments over an extended period of time. The intensity of sound around airports, along assembly lines of many plants, and in the operator's position of a pneumatic drill, a snowmobile, or a motorcycle all are examples of such noise environments. The use of high powered electronic amplifiers by "rock" bands provides "music" of an intensity which most certainly will lead to premature hearing loss of many of the young people who play in the bands and who enjoy listening to this form of music for long periods of time.

Our health and sense of well-being can be affected in other ways by noise which interferes with our sleep, our work, and our recreation. Many of us who have lived in modern apartment buildings lacking proper sound isolation or acoustical treatment have been irritated by a neighbor's air conditioning unit, TV, or other appliance which keeps us awake at night. Even the noisy ballast of a fluorescent lamp is bothersome in a quiet room when we are trying to read or make calculations. The noise from a canning factory or a drop forging plant or the hum of an electrical power transformer can be a source of annoyance to neighboring residential areas.

Noise which makes other sounds that communicate information to us can be bothersome, and it may even endanger our lives. For example, serious accidents can result when the din of a power lawnmower drowns out the warning sound of a horn as we step onto a busy street, or when the rush of air from a large fan in a factory muffles the warning shout of a coworker. People working underground in mines are very sensitive to the sudden sound of bursting or falling rocks, which forewarn them of impending danger. For this reason, it is very difficult to get miners who operate noisy equipment to wear protective earplugs, even though these plugs may be designed to let desired sound through. On the other hand, it may be difficult to sell a "quiet" vacuum cleaner, because the operator associates noise with power to suck up dirt from a floor.

Three adverse effects of noise on human being have been demonstrated in the examples given above:(1) hearing loss, (2) annoyance, and (3) speech interference. our ultimate goal in this study will be to learn to measure and evaluate a given noise environment based on specific criteria which relate to these three effects on people, and to apply standard noise control techniques to problem environments in order to bring them up to acceptable standards. The remainder of this chapter will be devoted to laying to groundwork needed to achieve these goals. We will develop the basic concepts of sound, define terms and build the vocabulary needed to describe noise problems, and specify and quantify noise criteria related to the effects of noise on humans.

New Words and Expressions

noise	*n.* 喧闹声，杂声，响声，不寻常的声音，噪声
interpretation	*n.* 理解，阐明，翻译，表演，演奏，解释
appreciation	*n.* 欣赏，赏识，正确评价，鉴别，感谢，增值
accrue	*vi.*（利息等）自然增长，出现
assembly	*vi.* 集合，调集，装配
pneumatic	*adj.* 空气的，气体的，装有气胎的

snowmobile　*n*. 履带式雪上汽车，乘雪上汽车旅行
motorcycle　*n*. 摩托车；*vi*. 骑摩托车，坐摩托车
high-powered　*adj*. 高功率的
amplifier　*n*. 放大器，扩音机
band　*n*. 团，队，伙，邦，带，波段，频带；*vt*. 用带子捆扎；*vi*. 团结
premature　*adj*. 早熟的，不成熟的，过早的
interfere　*vi*. 干涉，干预，妨碍，骚扰
recreation　*n*. 重建，再创作，娱乐
apartment　*n*. 房间，一套房间（英），一套公寓房间（美）
isolation　*n*. 隔离，孤立，脱离，分离
acoustical　*adj*. 听觉的，传音的，声学的
irritate　*vt*. 激怒，生气
ballast　*n*. 压舱物，镇流器；*vt*. 给……装压舱物
fluorescent　*adj*. 荧光的，发荧光的
bothersome　*adj*. 麻烦的，讨厌的
forging　*n*. 锻，锻件
hum　*vi*. 发哼哼声；*n*. 哼哼声，嘈杂声，嗡嗡声
endanger　*vi*. 危害，危及，使遭到危险
din　*n*. 闹声，嘈杂声；*vt*. 絮絮不休地说
lawnmower　*n*. 割草机
horn　*n*. 角，触角，角质，角质物，号角，喇叭；*vt*. 把……做成角状
muffle　*vt*. 消音，捂住；*n*. 消声器，马弗炉
coworker　*n*. 共同工作的人，合作者，同事
bursting　*n*. 爆炸，突发，一时迸发，突然破裂
forewarn　*vt*. 预先警告
impending　*adj*. 急迫的，紧急的
earplug　*n*.（防水或防震聋用的）耳塞
suck　*vt*. 吸，吸收，吸取；*vi*. 吸奶，吸；*n*. 吸
remainder　*n*. 剩余物，残余部分，剩下的人；*adj*. 剩余的
rock band　摇摆（滚）乐团（队）
specify　*vt*. 精确测定，拟定技术条件
electrical power transformer　电压变压器
fan　*n*. 风扇，鼓风机
drown out　淹没，压过

Reading Material 9

Urban Noise and Psychological Distress: A Systematic Review

Chronic exposure to urban noises is harmful to auditory perception, cardiovascular, gastrointestinal and nervous systems, while also causing psychological annoyance. Around 25% of the EU population experience a deterioration in the quality of life due to annoyance and about 5%-15% suffer from sleep disorders, with many disability-adjusted life years (DALYs) lost annually. This systematic review highlights the main sources of urban noise, the relevant principal clinical disorders and the most effected countries. This review included articles published on the major databases (PubMed, Cochrane Library, Scopus), using a combination of some keywords. The online search yielded 265 references; after selection, the authors have analyzed 54 articles (5 reviews and 49 original articles). From the analysis, among the sources of exposure, we found the majority of items dealing with airports and wind turbines, followed by roads and trains; the main disorders that were investigated in different populations dealt with annoyance and sleep disorders, sometimes associated with cardiovascular symptoms. Regarding countries, studies were published from all over the world with a slight prevalence from Western Europe. Considering these fundamental health consequences, research needs to be extended in such a way as to include new sources of noise and new technologies, to ensure a health promotion system and to reduce the risk of residents being exposed.

Noise pollution is defined as "noise in the living environment or in the external environment such as to cause discomfort or disturbance to rest and human activities, danger to health, deterioration of ecosystems, material goods, monuments, the external environment or such as to interfere with use of the rooms themselves". This type of pollution can mainly result from vehicle traffic, railways, airports, constructions, industries, recreational activities, etc. Worldwide, many people are exposed to this risk factor and they can suffer the relative consequences. In Western Europe, at least one million healthy life years are lost per year. Actually, as many as 125 million European citizens are exposed to road traffic, which is above the average annual levels of 55 dB, however, these figures could actually be significantly higher. Such an exposure causes a perception of annoyance for 20 million inhabitants. In 8 million inhabitants, sleep disorders appear, which causes more than 40,000 hospitalizations. In addition, around 8000 children in Europe are believed to have difficulties in reading and with concentrating in areas where air traffic noise is close to school buildings. Prolonged exposure to noise can be harmful to the auditory perception, with the onset of perceptual hearing loss, and to other human systems, in particular the cardiovascular, gastro-enteric, and nervous systems; it can also cause psychological annoyance, defined by ISO/TS 15666:2003(E) through the expression "one person's individual adverse reaction to noise". Road traffic noise can lead to the development of cardiovascular and metabolic disease and possibly oncological disorders. Additionally, this exposure may increase the risk of weight gain, obesity and Type Ⅱ diabetes mellitus. Data on the possible development of oncological pathologies are still controversial; some studies on urban noise demonstrated a positive association between these exposures and breast cancer; on the other hand, other studies found no association. A case-control study carried out on women found no association between cancer and traffic or railway noise, but a positive association with aircraft-noise exposure. Prolonged negative feelings towards noise may increase the risk of more severe psychological problems. It has been shown, through very well documented subjective data, that annoyance and sleep disturbances are the most widespread reported disorders associated with environmental noise. Tiredness, headaches and other psychological conditions are also associated with noise in adult populations. Psychological distress has been recognized as a substantial public health problem and as a leading cause of morbidity and disability. It accounts for most of the community burden of poor mental health. It has been estimated that around 25% of the EU population experience a deterioration in quality of life due to annoyance, and about 5%～15% suffer from sleep disorders. In fact, according to WHO, most of lost disability-adjusted life years (DALYs) can be attributed to noise-induced sleep disturbance and annoyance. Because of this, the EU has issued directives on the subject. The 2002/49/CE Directive has the primary objective of avoiding, preventing or reducing the harmful effects of exposure to environmental noise, by determining the exposure to noise (by means of acoustic mapping), public information on noise' effects and the adoption of action plans. In addition, Legislative Decree 194/2005 implements the previous directive on the determination and management of environmental noise; it defines the procedures of competences for the installation of strategic noise maps in urban areas with more than 100,000 inhabitants, guaranteeing public participation. This systematic review aims to identify the sources of urban noise that cause the most discomfort to residents, the main psychological disorders associated with the condition and the countries which are most effected.

Considering the constantly growing trend of new sources of noise and the particular susceptibility of people, caused by numerous factors, it is becoming increasingly urgent to define the extent of noise exposure, its severity and the correlation between sound input and the deterioration of the quality of life caused in the population. In 2005, the European Commission dedicated the European Week on Workplace Health and Safety to noise, developing numerous information and communication initiatives aimed at raising public awareness of this risk agent. In order to address the problem of environmental noise with long-lasting solutions, it is therefore necessary to quantify the effects of external noise, either to predict new socio-economic impacts or in relation to the health of residents, to develop new policy strategies and finally, to create new guidelines. These should aim at easing the severity of the problem and avoiding complications in the medium to long term. In order to do this, it is clear that socio-acoustic survey is an indispensable tool for standardizing the correlation between noise reactivity and the extent of provocative noise.

写作技巧

正文的写作

正文是科技论文的主体部分，通常包括引言（introduction）、材料与方法（materials and method）、结果与讨论（results and discussion）和结论与建议（conclusion and suggestion）等方面。在论文的实际写作过程中，撰写各章节的先后顺序一般为methods—results—discussion—introduction—title—abstract。

1. 引言

引言用于说明论文写作的背景、理由、主要研究成果及其与前人工作的关系等，目的是引导读者进入论文的主题，并让读者对论文中将要阐述的内容有心理准备，因此，引言有总揽论文全局的重要性，也是论文中最难写的部分。

引言必须包含介绍研究背景、提出研究问题和阐述研究目的等内容。

2. 材料与方法

依据写作的论文属于实验型、理论型或描述型，“材料”有所不同。实验型的研究论文中使用的材料可以是设备、物料和试剂；理论型的研究论文可以是研究区域相关特征、指标或以人为研究手段获得的信息数据；描述型的研究论文的材料可以是调查问卷、样本等。

论文中应介绍研究采用的方法，主要有三个方面的原因：首先，所用方法是读者评估课题意义的重要手段。如果研究程序有误，研究结果的可靠性就值得怀疑。因此应以充分的理由说明所用方法既正确又可靠。其次，如果采用某种新方法或以非常规方式使用某种较陈旧的方法，介绍研究方法的重要性更是不言而喻。最后，科学研究中有一条法则，凡采用同一方法，应该可以取得相同的结果，因此，为使他人能取得相同的结果，必须详细描述课题研究中采用的方法。

大多数作者在阐述过程与方法时，最常见的问题是泛泛而谈、空洞无物，只有定性的描述，使读者很难清楚地了解论文中解决问题的过程和方法。因此，在说明过程与方法时，应结合（指向）论文中的公式、实验框图等来进行阐述，这样可以既给读者一个清晰的思路，又给那些看不懂中文（但却可以看懂公式、图、表等）的英文读者以一种可信的感觉。

3. 结果与讨论

介绍完材料与方法之后应该接着介绍作者的研究结果，研究结果是遵循论文描述的研究程序取得的，因此作者在撰写这两部分时应反映两者之间的联系。

研究结果是论文的关键，论文应对其研究结果予以充分说明。发表研究结果的最终目的是使研究成果为读者所了解，通常可分三步逐一介绍。

- 先以一个句子作概括性介绍，例如：

The results obtained indicate that the film temperature increases as the rolling velocity increases.

- 列出代表性的数据或分项目。

除代表性数据外，如果还有其他数据需要发表，这类数据可列入附录，不必列入正文中。

如果有大量数据需要列出，应该使用图表。图表是指曲线圈、示意图、流程图、图画、照片、表格等，使用各类图表不仅可以使论文主要内容易被读者了解，而且省时有效。论文的文字部分可以指出图表中的数据，但不必用文字将数据再重复一遍。图表是科技文献的特征之一，图表必须简明、扼要、形象、具体，并具有鲜明的对比性。使读者一望而知全局及其与左右前后的关系。

- 研究结果和对研究结果的分析判断，通常要逐项探讨。

如果有些研究结果在某些方面出现异常，作者无法解释，即便不影响论文的主要论点，也应在论文中加以说明。这种情况不一定表示研究工作失败，因为在变革自然的实践过程中，人们原定的思想、理论、计划和方案很少能毫无改变地实现出来。指出存在的异常现象有助于自己和他人今后继续研究这些异常现象。历史上许多重要的科学发展就是反复研究异常现象的结果。

为了帮助读者更好地理解研究结果的意义，作者应对研究结果进行适当的评述。这种评述可通过四种不同的方式进行：叙述事实、介绍一定时空范围内的变化过程、进行对比分析和进行因果分析。

① 叙述事实。例如 It was verified that the average error in fuel spray droplet size is less than 10%。

② 介绍变化过程。例如 The interface between the saturated and unsaturated regions existed, but did not continue to exist when the water content was increased.

③ 进行对比分析。例如 Our result is not agreement with Newton(1685).

④ 进行因果分析。例如 Obviously, the rise in temperature causes the expansion of the gas.

4. 结论与建议

结论是论文的总观点，是研究结果的逻辑发展，是整篇论文的归属。许多读者喜欢阅读研究结论甚于研究过程本身，因此，结论必须完整、准确、鲜明。

罗列研究结果不能代替结论。结论比研究结果和分析更进一步，要反映作者如何将研究结果经过概括、判断和推理而形成总的观点，要反映事物的内在联系。

结论与建议部分有必要提出如何深入研究有关问题，可以对自己提出这样的问题：如果还有机会再继续做这一课题，该做什么？作者在该课题上已投入很多时间，无疑比别人更清楚该课题还有哪些尚未做、值得自己和别人继续做的工作。

如果没有确切的结论或建议，不要勉强杜撰。但不要错过或遗漏任何真正的结论，因为凡是有价值的研究工作都能在一定程度上完成某一具体过程的认识活动，得出或证实一条或几条有意义的结论。

Exercises

Ⅰ. Point out true or false.

1. Noise has many adverse effects on human beings, such as hearing loss, disrupting communication, and even hampering thinking ability.

2. Noise interferes with our work, study and living.

3. Noise can be controlled by three ways: source of noise, receiving body of noise and transport path.

4. The word “noise” implies the response of a human being to a noise environment, so it also implies a value judgment of the sound.

5. Noise from “rock” bands' performances is not likely to cause hearing loss.

6. Noise from machines in factories can never be bothersome to nearby residential areas.

7. Protective earplugs are widely used by miners operating noisy equipment.

8. The study aims to evaluate noise environments based on their effects on human health.

9. Annoyance and speech interference are not considered adverse effects of noise.

10. The study in the text focuses only on measuring and evaluating noise levels.

Ⅱ. Answer the following questions according to the text.

1. What will happen if you live in a building lacking proper sound isolation or

acoustical treatment?

2. What adverse effects of noise on human are there?

3. What is the primary meaning of the word "noise" as discussed in the text?

4. How can exposure to moderate noise environments over time lead to hearing damage?

5. Give examples of noise environments that can interfere with sleep, work, and recreation.

Ⅲ. Scan the QR code to read the article "China Continues to Tackle Noise Pollution in 2021—2025", and then discuss how can technological advances be harnessed to improve the detection and mitigation of noise disturbances, thereby promoting substainable coexistence between communities and the surrounding natural environment?

Unit 10

Soil Pollution and Remediation

【重点与难点】

（1）科技文献的标注方法；（2）科技论文中致谢部分的几种常用句型；（3）快速阅读的方法。

【学习指南】

依据课程重难点，结合课程线上资源及教材的课后习题完成课前预习和课后复习，掌握科技文献快速阅读的方法，自觉应用辩证思维解决土壤修复瓶颈。

Text 10

Techniques for Removal Pollutants from Soil

Polluted soil can be cleaned in very many different ways. Of course, the best way is to prevent this pollution, but unfortunately we have to deal with an heritage of the past generation as well as accidents that may cause polluted areas. Therefore cleansing operations will be necessary now and in the nearby future.

There are as many sanitation techniques as there are polluted sites. The best technique one has to apply will be dependent on properties of the polluting compounds as well as the properties of the soil. Maybe some external factors play a role in the choice as well: limiting financing, unacceptable risks for the neighborhood, limiting time, (industrial) activities upon the site, future activities that are planned, the (future) use of the soil. So it will be necessary to have a thorough knowledge of soil properties, the interference between soil, technique and compound.

In general it can be said, that all techniques can be divided into two categories: in situ-techniques, that are performed without excavating the soil, and ex situ-techniques, that are performed after the soil has been removed from the site, and transported to, most often-large cleansing plants, where the polluted soil will be treated.

A soil distillation process that can be for a wide range of pollutants and soils, as either as a discrete unit or in combination with a washing plant, has now been developed by Harbauer India, an environment technology company. Fine-grain and highly contaminated soils, which so far have been particularly problematic, can be treated with the help of this process.

The basic principles are described below:

Biological and thermal processes already exist for the treatment of such fraction. Biological processes can be applied to only a limited extent on account of the low degradability of organic residues that are obtained. Metal compounds cannot be removed from contaminated soils using conventional thermal processes. Thermal processed (incineration) are well suited to be the cleaning of soil contaminated by purely organic pollutants. However, thermal treatment of very fine-grain particles involves costly and complex technological processes. The substantial volumes of flue gases produced by conventional thermal processes for soil decontamination are a significant drawback. This leads to high investment costs for cleaning processes.

The range of application for vacuum distillation is limited, however, to pollutant compound with defined boiling points. These include a large number of organic compound, among them crude oil fraction, halogenated hydrocarbon, PAHs, cyanides and metal compound that vaporize.

The vacuum distillation process involves heating the soil under a vacuum to a predefined temperature. The pollutants evaporate and are then transported to a condenser using inert gas. In the condenser, the contaminants are condensed and then extracted. The condensates are either treated for possible re-use or are sent to a sophisticated station incinerator for disposal. Non-combustible pollutants must be disposed of properly. Surplus inert gas and air mixture is either fed back into the cycles or is released into the atmosphere via a small-scale waste gas purification plant.

The optimal temperature is determined in preparatory experiments. The desired temperature should be as low as possible in order to prevent the pollutant compounds from decomposing. The boiling points of the various pollutant compounds can be influenced by varying the pressure.

Soil distillation is an economic solution for the cleaning of fine-grain, high-contaminated soils or fine-grain residues obtained from soil cleaning plants. The significant benefit of this process is the low volume of process gas required. This, in turn, means low investment costs and relatively quick official approval.

Contaminants are extracted by this process in the lowest possible volumes and in highly concentrated form, and can either be treated for reutilization or must be disposed of in the correct manner. Fixed incineration plants have a high level of technical sophistication and the full range of waste gas purification equipment is suitable for disposing of organic pollutants. This guarantees that the contaminants removed from the soil are destroyed in plants that correspond to the current state of technological development.

New Words and Expressions

heritage	*n.* 遗产，继承权，传统
sanitation	*n.* 卫生，卫生设施，卫生设备
excavate	*vt.* 挖掘，开凿，挖出，挖空，掘出
distillation	*n.* 蒸馏，蒸馏法，蒸馏物
discrete	*adj.* 分离的，无关联的，不连续的，离散的
thermal	*adj.* 热的，热量的，温泉的；*n.* 上升的热气流
flue	*n.* 烟洞，烟道，暖气管
halogenate	*vt.* 卤化
cyanide	*n.* 氰化物
combustible	*adj.* 可燃的，易燃的；*n.* 可燃物，易燃物

Reading Material 10

Review of Soil Heavy Metal Pollution in China: Spatial Distribution, Primary Sources, and Remediation Alternatives

As an important part of the terrestrial ecosystem on earth, soil is the most basic natural resource for human being living on, which could constantly regenerate and recycle. Since the beginning of the 20th century, soil heavy metal pollution has become an intractable environmental problem, which not only affects substance exchange as well as energy conversion, also causes soil resource degradation and depletion, in turn causes ecosystem deterioration. Thus, the rational utilization of soil resources and the protection of soil environment are concerned all over the world.

Heavy metals in soil can originate from a wide range of sources, including weathered parent materials, effects of mining, traffic emissions, products of smelting, application of chemical fertilizers, and other industrial, agricultural, and business activities (Jin et al.,

2019). Soil heavy metal pollution not only damages soil quality (Hu et al., 2020) and reduces crop yields (Yang et al., 2018a), also jeopardizes water and atmospheric environment, threatens human health, aggravates global climate change, and affects the sustainable development of society (Briffa et al., 2020). Therefore, soil heavy metal pollution has attracted global attention and has been listed as a priority for pollution monitoring and controlling (Lacarce et al., 2012; Schneider et al., 2016; Toth et al., 2016; Lequy et al., 2017; Marchant et al., 2017; Ye et al., 2017, 2019).

With rapid industrialization and urbanization in recent decades, heavy metal contamination in soil has become serious and widespread across China (Duan et al., 2016; Zhang et al., 2016; Zhong et al., 2018). Most soil pollutants are inorganic pollutants, 82.8% of which exceed the national standard level (Department of Natural Ecology Protection, Ministry of Environmental Protection, 2012). According to the Bulletin of the National Survey of Soil Pollution, the heavy metals Cd, Hg, As, Cu, Pb and Zn exceed the standards of 7.0%, 1.6%, 2.7%, 2.1%, 1.5% and 0.9%, respectively (Ministry of Ecology and Environment of the People's Republic of China, 2014). Previous studies have reported the phenomenon of heavy metal pollution in some heavily polluted regions, such as the old industrial bases in Northeast China (Chai et al., 2014; Li et al., 2014; Zong et al., 2017), the Yangtze River Delta (Cui et al., 2004; Hu et al., 2019), the Pearl River Delta (Zhang et al., 2018), and central regions, for example, Hunan and Hubei Province (Ding et al., 2017; Chen et al., 2018). However, only some studies in China have focused on heavy metal pollution at the national scale (Zhang et al., 2018; Zhou et al., 2018).

Previous studies indicated that the main sources of soil heavy metal pollution in China comprise natural factors (such as topography, climate, etc.) (Fu et al., 2018b) and anthropogenic factors (such as farming, industry, etc.) (Minasny et al., 2010). Although a large number of research has been carried out on the driving factors of soil pollution, most of which are still broad studies of industry, agriculture, urban, etc (Dazzi et al., 2009; Lo Papa et al., 2011; Shangguan et al., 2014; Dazzi and Giuseppe, 2016; Fu et al., 2019). More discussions are needed to focus on specific driving factors, including human behaviors e.g. domestic pollution sources inducing soil heavy metal pollution. Furthermore, main remediation measures of soil heavy metal pollution were discussed, such as physical, chemical and biological remediation technologies (Audrone et al., 2005; Han et al., 2019; Shi et al., 2020; Singh et al., 2020). However, systematic research on remediation technologies for soil heavy metal contamination is rare as the practical cases of remediation technology applications. These limitations suggest that a comprehensive survey of soil heavy metal pollution including spatial distribution and primary sources, as well as remediation measures in China is required to implement specific and effective management.

In this study, we collected more than 2000, 1600, 1300, and 1200 sampling points of soils, which are typical areas where Cd, Pb, Zn, and Cu occur in excess, respectively, across China, based on previously published research studies. Based on these data, we investigated the geographical distribution of heavy metal pollution and potential dominant environmental and urban development factors that caused the pollution. The specific purposes of our research were ① to explore the pollution status and spatial features of soil heavy metal contamination in recent 10 years; ② to explore the main factors that cause soil heavy metal pollution in representative cities; and ③ to compare and analyze which remediation technologies (physical, chemical, biological, or combined methods) are more suitable for different contaminated sites. The results of this study will enhance our understanding of the status of soil heavy metal pollution (mainly Cd, Pb, Zn, and Cu) in China and provide valuable information for the targeted remediation of affected soils, so as to ensure proper development and utilization of soil resources.

We systematically studied the current status of soil heavy metal pollution across China, including pollution positions, primary causes for the pollution and alternative remediation techniques. Main views and prospects are provided as follows:

(1) Soil heavy metal pollution in southern China is more serious than that in northern China. Soil heavy metals (Cd, Pb, Zn, and Cu) seriously exceed their standard levels. In Northeast, Central, Southwest and South China, the contents of Cd and Pb increased gradually from northwest to southeast and from northeast to southwest. In terms of the 4 heavy metals studied in this paper, Cd pollution is the most serious in China, followed by Pb, Zn and Cu pollution.

(2) The severity of soil heavy metal pollution varies among different provinces. Combined heavy metal pollution in soils occurs in Northeast China (represented by Liaoning Province) and in southern China (represented by Guangxi and Yunnan Province). Mineral exploitation could be the primary factor that has led to heavy metal pollution in Liaoning Province, while in Guangxi and Yunnan Province, industrial processing (e.g., mineral separation and coal preparation) have brought both opportunities and environmental challenges. Furthermore, East China (including Fujian and Shanghai Province), Central China (including Henan, Hubei and Hunan Province) and North China (including Hebei and Gansu Province) experiences single or multiple heavy metal pollutants in their soils. Multiple factors explain soil heavy metal pollution. Among them, mineral exploitation and industrial production are the primary causes of heavy metal pollution in the most severely polluted provinces, followed by sewage irrigation and

fertilizer application, which is reflected in some predominantly agricultural provinces. Urban development is the third main factor, especially in densely populated provinces. In addition, daily activities (including those related to transportation and household refuse) could be other reasons.

(3) An increasing number of areas polluted with heavy metals areas are using physical, chemical and biological combined remediation technologies. These technologies not only improve the remediation efficiency of contaminated soil but also overcome the limitations of a single remediation technology. In heavily polluted soil (source of heavy industry), chemical-physical combined remediation technologies are mostly adopted, such as soil replacement solidification/stabilization technology, solidification/ stabilization-leaching technology, application of chelating agent leaching technology. For moderately or lightly polluted soils (irrational agricultural practices or human activities), biological remediation combined with other technologies or various bioremediation technologies (e.g., phytoremediation -solidification/ stabilization, phytoremediation-microbial remediation and microbial-vermiremediation-phytoremediation) are jointly applied. Remediation technologies should be selected in accordance with site conditions, pollution features and remediation effects.

(4) Although the development of remediation technologies is remarkable, there is still room for improvement. First, advanced repair materials and equipment need to be developed. Second, the effective remediation methods mentioned above should be extended at a national scale. Third, it is necessary to change from ex situ remediation to in situ remediation, from fixed equipment to mobile equipment, from physicochemical methods to bioremediation based on combined remediation technologies, and from pollution remediation to pollution prevention.

(5) Moreover, China started to formulate a series of policies on environmental risk assessment and control since 2014, including the Guidelines for Soil Remediation of Contaminated Sites (HJ 25.4-2014) (Ministry of Environmental Protection of the People's Republic of China, 2014a), the Guidelines for Risk Assessment of Contaminated Sites (HJ 25.3-2014) (Ministry of Environmental Protection of the People's Republic of China, 2014b), Soil Environmental Quality: Risk Control Standard for Soil Contamination of Development Land (GB 36600-2018), and Soil Environmental Quality: Risk Control Standard for Soil Contamination of Agricultural Land (GB 15618-2018) (Ministry of Environmental Protection of the People's Republic of China, 2018a, 2018b). Since the National Action Plan for Prevention and Control of Soil Pollution introduced in May 2016, sources control and risk prevention of soil heavy metal pollution should be widely

concerned by the whole society, together with pollution remediation. Optimizing environmental management and improving the remediation efficiency of combined remediation technologies are equally important for soil environmental protection.

致谢和参考文献

1. 致谢

论文的致谢部分是对给予支持和帮助的单位和个人表示感谢，致谢时应说明其贡献和责任。致谢部分常由以下内容组成：

- 对为研究工作提供方便和帮助的实验室或个人表示感谢，尤其是对为研究工作提供实验设备或专门材料的人表示感谢。
- 对外来的经济上的支持者表示感谢。例如，对以赠送、合同或研究基金等形式为作者提供计划外的财政援助的个人或机构予以致谢。
- 对为本研究工作出主意、提建议、帮助解释实验现象的个人，可如实指出其具体内容，说明他们所起的作用。

英语科技论文中致谢部分的例文如下：

The authors greatly acknowledge Dr. Sudhakar, Head, RRSSC and other scientists of the center for their encouragement and technical guidance in carrying out this research. The authors also thank Soil Conservation Dept ., DVC, Hazaribagh for providing valuable data for validation of the simulation results.

作者感谢 RRSSC 负责人苏德哈卡博士以及该中心的其他科学家在本研究中的鼓励和技术指导。作者也感谢 DVC 水土保持部的赫泽里弗提供了验证模拟结果的有用资料。

英语科技论文中致谢部分常用句型：

- 项目资助

  The work (upon which this paper is based) was sponsored by …（机构）

  Through the project of …

  Support for this work from the …（机构）and … funds.

  Some / all of this work was funded through the …

  This article was funded by …
- 个人帮助

  The authors greatly acknowledge …（某人）of …（单位）for his/her encouragement and technical guidance in carrying out this research.

This article is dedicated to …

Thanks are due to (someone / establishment) for (something).

The authors are thankful to (someone /establishment) for (something).

We are grateful for the technical assistance of someone.

- 实验工作

…（某项实验）were/was performed under the supervision of someone.

…（某项实验）were/was developed at … laboratory.

2. 参考文献

科技论文列举参考文献是传统惯例，其反映作者严肃认真的科学态度并提供研究工作的广泛依据。论文写作中，凡引用其他作者的文章、观点或研究成果，都应该在文中标明（文中标注），并在参考文献中说明出处（文后著录）。文中标注和文后著录常有配套要求，共同构成著录整体。目前各类期刊在著录编排上并不统一，因此，如果欲在某期刊发表论文就应了解该期刊对著录格式的要求。现介绍常见的参考文献编排格式。

（1）文中标注的格式

文中标注有两种格式，哈佛标注制和顺序编码标注制。

- 哈佛标注制

正文中涉及文献引用时，在圆括号内标明引用文献作者的姓名和出版年。如果论文中引用了同一作者在同一年份发表的多篇文献，应按论文中引用文献的先后顺序，在出版年份后面加小写英文字母 a 、b 、c 等，以示区别。还有一种情况是被引用的作者姓氏已在文中列出，圆括号内就只标明出版年。例如：

Numerous lumped and distributed parameter，H/WQ models, including CREAMS (Knisel, 1980), ANSWERS (Beasley and Huggins, 1982), AGNPS (Young, 1989) and SWRRB-WQ (Arnold et al. , 1990), have been developed to predict the impacts of agriculture on the quality of surface water.

Satellite imagery has been widely used in the hydrologic modeling (Baker, 199la). Satellite remote sensing is an excellent tool for environmental monitoring, as it allows repeated coverage of areas on a regular basis (Baker, 1991b).

A comprehensive description of the AGNPS model can be found in Young et al. (1959).

- 顺序编码标注制

参考文献按论文中引用的先后顺序从 1 开始连续编号。正文中的编号一般用方括号括起来，放在引用文献的作者姓名或引用概念之后。如果同一个地方引用多篇文献，只需将各篇文献的序号在方括号内全部列出，各序号间用逗号隔开。如果遇到连续序号，可用符号“-”连接，略去中间序号。如果需要为整段内容注明参考文献，编号应放在此段最后一句上。目前，顺序编码标注制在国内外科技期刊与科技著作中使用频率最高。例如：

A method of computing LS for irregular slopes was published in 1975[1].

Respectively, Beasley and Huggins estimated the soil loss from Morrow lake watershed of Michigan in 1991[1-2].

Herbicides in water samples were extracted with CH_2Cl_2 and quantified by gas chromatography/mass spectrometry (GC/MS) using selected monitoring[20,23-26].

（2）著录的格式

作者或编者为3人以下的（含3人），应全列出；为3人以上的，在第3作者或编者后加“，等”。每种参考文献的著录项必须完整（包括用方括号标注文献类型）。

期刊论文：[序号] 作者. 题名[J]. 刊名（外文也要写出全称，不缩写），出版年，卷号（期号）：页码.

专　　著：[序号] 作者. 书名[M]. 版本（初版不写）. 出版地：出版者，出版年：页码.

论 文 集：[序号] 作者. 题名[C]// 文集编者. 文集名. 版本. 出版地：出版者，出版年：页码.

学位论文：[序号] 作者. 题名[D]. 论文保存地：论文保存单位，论文完成年：页码.

报纸文章：[序号] 题名[N]. 报纸名，出版日期(版次).

国家标准：[序号] 标准编号，标准名称 [S].

专　　利：[序号] 专利所有者. 专利题名：专利号[P]. 公告日期.

电子文献：[序号] 作者. 电子文献题名[电子文献及载体类型标识]. 电子文献的出处或可获得地址，发表或更新日期/引用日期（任选）.

未定义类型的文献：[序号] 作者. 文献题名[Z]. 出版地：出版者，出版年.

著录规则举例如下：

[1] 陈怀满. 土壤中化学物质的行为与环境质量[M]. 北京：科学出版社，2002：45-47.

[2] 夏家淇，骆永明. 我国土壤环境质量研究几个值得探讨的问题[J]. 生态与农村环境学报，2007，23(1):1-6

[3] Perez D M A, Burgos P, Madejon E, et al. Microbial community structure and function in a soil contaminated by heavy metals: effects of plant growth and different amendments [J]. Soil biology and biochemistry, 2006, 38:327-341.

[4] 胡文. 土壤-植物系统中重金属的生物有效性及其影响因素的研究[D]. 北京：北京林业大学，2009：12-14.

[5] Dressing, Steven A. Non-point source management system software [EB/OL]. http:// www. epa. gov/ owow/ watershed / proceed / dressing.htm, 1999-04-10.

Exercises

Ⅰ. Point out true or false.

1. The article mainly discusses how to reduce the cleansing cost.

2. The best technique to apply to solve pollution will be dependent on both properties of the pollution compounds and soil.

3. Fine-grain and highly contaminated soil can be treated with the help of biological process.

4. Biological processes can be applied to only a limited extent because of the low degradability of organic residues that are needed.

5. Organic residues cannot be removed from contaminated soils using conventional thermal processes.

6. The principal benefits of the vacuum distillation process are as follows except that oxygen concentration of process is high.

7. Heating under vacuum is an economic solution for the cleaning of fine-grain, high-contaminated soils or fine-grain residues obtained from soil cleaning plants.

8. The boiling points of pollutant compounds are not affected by pressure variations.

9. Non-combustible pollutants can be released into the atmosphere without any treatment.

10. Fixed incineration plants are not equipped to handle organic pollutants from soil cleaning.

Ⅱ. Translate the following sentences.

1. 项目来源于国家自然科学基金。

2. 我的妻子的支持

3. David 教授的技术指导和鼓励

4. 特别感谢 XYZ 实验室为我们提供必要的实验设备和资源。

5. 我要感谢我的家人和朋友，在这段充满挑战的旅程中给予我坚定的鼓励和理解。

Ⅲ. Mark the reference

一本名叫“soil and water conservation”期刊，文章名称“chemical and bacteriological quality of pasture runoff”，作者是 Doran J W，期刊出版年是 2000 年，第 20 卷第 2 期。参考内容在 34-36 页。

Ⅳ. Scan the QR code to read the article "Draft Law to Help Protect Black Soil" and then discuss what are the technical difficulties in soil pollution remediation and treatment, and how to solve the problem of meeting the current product demand and the long-term goal of sustainable soil health and ecological protection?

Unit 11

Modern Instrumental Analysis

【重点与难点】

（1）常见现代仪器分析技术；（2）水质分析技术发展现状；（3）科技论文投稿步骤及注意事项。

【学习指南】

依据课程重难点，结合课程线上资源及教材的课后习题完成课前预习和课后复习，了解科技论文投稿步骤及注意事项，自觉应用马克思主义哲学思想和方法论解决环境样品分析过程中设备选型的问题。

Text 11

Modern Instruments for Environmental Sample Analysis

Modern instrumental analysis technology has revolutionized environmental sample analysis, providing rapid, sensitive, and accurate methods for detecting and quantifying various pollutants and compounds in environmental samples. These techniques help researchers, industries, and policymakers monitor and mitigate pollution and understand the impacts of human activities on the environment. Some of the most widely used modern instrumental analysis technologies in environmental sample analysis include:

Gas chromatography-mass spectrometry (GC-MS): This technique is extensively used to analyze volatile organic compounds (VOCs) in air, water, and soil samples. GC-MS combines gas chromatography, which separates complex mixtures into individual components, and mass spectrometry, which identifies and quantifies the components based on their mass-to-charge ratio.

Liquid chromatography-mass spectrometry (LC-MS): LC-MS is suitable for analyzing non-volatile, thermally labile, and polar compounds in environmental samples, such as pesticides, pharmaceuticals, and personal care products. Like GC-MS, it separates components using liquid chromatography and identifies and quantifies them using mass spectrometry.

Inductively coupled plasma mass spectrometry (ICP-MS): ICP-MS is a powerful technique for trace elemental analysis in environmental samples, such as water, soil, and air particulates. It is capable of detecting and quantifying elements at extremely low concentrations (parts per trillion levels) with high accuracy and precision.

Atomic absorption spectrometry (AAS): AAS can be used for the analysis of water, soil and other samples, and can detect the content of metal elements such as lead, mercury and other heavy metals. It has the advantages of high sensitivity and high accuracy, which can accurately measure and analyze trace metal elements, and play an important role in environmental pollution monitoring and control.

X-ray fluorescence (XRF): XRF is a non-destructive technique used for elemental analysis of solid environmental samples, such as rocks, soils, and sediments. It provides information about the elemental composition of the samples by measuring the characteristic X-rays emitted by the elements when they are excited by an X-ray source.

Fourier transform infrared spectroscopy (FTIR): FTIR is a non-destructive analytical technique that measures the infrared absorption spectra of environmental samples to identify and quantify various organic and inorganic compounds. It is widely used to analyze air, water, and soil samples for pollutants such as VOCs, greenhouse gases, and particulate matter.

Nuclear magnetic resonance (NMR) spectroscopy: NMR is a powerful technique for characterizing the molecular structure and composition of organic compounds in environmental samples. It provides information about the molecular structure, functional groups, and chemical environment of the compounds in the sample.

Remote sensing: Remote sensing techniques, such as satellite imagery and LIDAR, enable large-scale monitoring of environmental parameters, including land use/land cover changes, vegetation health, and water quality. These data can be used to assess the impact of pollution, climate change, and other environmental stressors on ecosystems.

These modern instrumental analysis technologies have significantly improved the speed, sensitivity, and accuracy of environmental sample analysis. By providing reliable and comprehensive data on the presence and levels of pollutants in the environment, they support informed decision-making for pollution prevention, mitigation, and remediation efforts.

New Words and Expressions

chromatography	*n.* 色谱分析法
spectrometry	*n.* 光谱测定法
volatile	*adj.* 易变的，易挥发的，易气化的；*n.* 挥发物
inductively	*adv.* 归纳地，诱导地
accuracy	*n.* 准确性，精确性
precision	*n.* 精确（性），准确（性）；*adj.* 精密的，精确的
fluorescence	*n.* 荧光，荧光性
non-destructive	*adj.* 无损的，非破坏性
particulate	*adj.* 微粒的；*n.* 微粒，微粒状物质
parameter	*n.* 界限，范围，参数，变量

Reading Material 11

Surface Enhanced Raman Spectroscopy in Environmental Analysis, Monitoring and Assessment

Environmental pollution is usually monitored via mass spectrometry-based approaches. Such techniques are extremely sensitive but have several disadvantages. The instruments themselves are expensive, require specialized training to use and usually cannot be taken into the field. Samples also usually require extensive pre-treatment prior to analysis which can affect the final result. The development of analytical methods that matched the sensitively of mass spectrometry but that could be deployed in the field and require minimal sample processing would be highly advantageous for environmental monitoring. One method that may meet these criteria is Surface Enhanced Raman Spectroscopy (SERS). This is a surface-sensitive technique that enhances Raman scattering by molecules adsorbed on rough nanostructure surfaces such as gold or silver nanoparticles. SERS gives selective spectral enhancement such that increases in sensitivity of 10^{10} to 10^{14} have been reported. While this means SERS is, theoretically at least, capable of single molecule detection such a signal enhancement is hard to achieve in practice. In this review, the background of SERS is introduced for the environmental scientists and the recent literature on the detection of several classes of environmental pollutants using this technique is discussed. For heavy metals, the lowest limit of detection reported was 0.45 μg/L for mercury; for pharmaceuticals, 2.4 μg/L for propranolol; for endocrine disruptors, 0.35 μg/L for 17β-estradiol; for perfluorinated compounds, 500 μg/L for perfluorooctanoic acid and for inorganic pollutants, 37g/L for general pesticide markers. The signal enhancements achieved in each case show great promise for the detection of pollutants at environmentally relevant concentrations and, although it does not yet routinely match the sensitivity of mass spectrometry. Further work to develop SERS methods and apply them for the detection of contaminants could be of wide benefit for environmental science.

The traditional (and very effective) method to detect micropollutants in the environment is to use chromatography as a separation step followed by mass spectrometry (ideally high-resolution) for detection. Liquid chromatography – mass spectrometry (LC-MS) and gas chromatography – mass spectrometry (GC–MS) are the two most widely used techniques. Both GC–MS and LC–MS have high sensitivity (down to ng/L levels or lower) but the instruments themselves are complex and require a high degree of specialized training to use and maintain. Extensive sample pre-treatment is also usually required prior to analysis. This means that both GC–MS and LC–MS are costly and time consuming to use for high throughput analysis and unsuitable for realtime monitoring in the environment. A method that could detect target compounds faster and with minimal sample preparation would thus be highly advantageous. One such method is Raman Spectroscopy. Although by itself this technique cannot match mass spectrometry for sensitivity the use of Surface Enhanced Raman Spectroscopy (or SERS) may allow circumvention of this problem.

Briefly, SERS is a spectroscopic technique that relies on electronic and chemical interactions between the excitation laser of the spectrometer, the analyte of interest, and a particular substrate to selectively boost the signal, and thus detection, of target molecules. Among the wide variety of available SERS substrates, colloidal metal (usually gold-or silver based) nanoparticle systems are the most widely used due to their effectiveness, ease of preparation of the nanoparticles, and the ability to tune analytic sensitivity through chemically-controlled variation of nanoparticle type and size (Tian et al., 2014). Raman Spectroscopy requires comparatively little pre-treatment of samples and is quick to perform, with measurement times being on the scale of seconds to minutes compared to tens of minutes (and in some cases hours) for conventional, chromatography-mass spectrometry-based methods (Jones et al., 2003). This means SERS is able to boost signal detection by as much as 10^{10} or 10^{14} in a similar time frame as standard Raman, with single molecule detection being theoretically possible. In practice, single molecule detection is very hard to perform outside very specific laboratory set ups but SERS is still increasingly widely used for its high sensitivity, fast analysis time (with minimal sample preparation), and the fact that it is amenable to many different target compounds, including pollutants.

SERS can potentially not only save a significant amount of time and money in environmental detection but also allow remote and automatic sampling of water sources on a large scale (Halvorson and Vikesland, 2010). Several obstacles remain to be overcome before this can become routine, however, not least of which is how to improve reproducibility of measurements at low concentrations from real world samples as opposed to lab studies.

This review aims to provide an overview of the current state of SERS in the field of environmental pollutant detection. The contaminant types covered are heavy metals, pharmaceuticals, hormonal and endocrine disrupting compounds, perfluorinated compounds (such as Perand Polyfluoroalkyl Substances -PFAS), and pesticides. An assessment of current strengths and weaknesses in using SERS for environmental research and recommendations for further research are also presented.

Raman spectroscopy boasts much higher efficiency than mass spectrometry in terms of sample pre-processing and measurement runtime. Measurements can be run in the field, an ability most mass spectrometry-based techniques lack. Crucially, when combined with surfaced enhancement techniques, Raman spectroscopy can potentially match mass spectrometry in terms of sensitivity. The development of this latter point is however, quite variable at present and this is where the bulk of the current research is focused.

The groundwork has been done for developing SERS sensors to provide a sufficient Raman signal increase to detect a target analyte at environmentally relevant concentrations. However, this has mainly (but not exclusively) been tested in the lab using benchtop Raman spectrometers, perhaps with wireless capability for data processing. The next step is to adapt the method and SERS systems to be used in the field. Portable Raman spectroscopy and SERS have the potential to meet the demands set for performing real-time environmental water quality monitoring in the field, at concentrations low enough to detect pollutants at the concentrations that are thought to exist in the environment itself. Complex mixtures of compounds can potentially be dealt with via the development and use of spectral libraries of common pollutants, while methodologies need to be optimised before the technique could become widespread. There are an increasing number of studies that have used SERS to detect pollutants in the environment. The challenge for the future is to develop and raise awareness of the potential benefits of SERS to environmental analyses.

科技论文的投稿

1. 投稿步骤

科技论文的投稿过程涉及多个步骤，作者需要在每个阶段都投入时间和精力以确保论文的质量和准确性，主要步骤如下：

- 选定目标期刊：首先，要确定一个与本人研究领域相关的合适期刊。仔细阅读期刊的投稿指南，确保论文满足期刊的格式和内容要求。
- 撰写论文：按照期刊的投稿要求和规定格式编写论文，包括摘要、引言、方法、结果、讨论、结论、参考文献等部分。
- 同行评审：在提交论文之前，让同行或导师审阅，以获取关于论文质量、可读性和准确性的反馈。根据评审者的建议进行必要的修改。
- 准备附件：根据期刊的投稿要求，准备必要的附件，例如封面信、作者声明、利益冲突声明等。确保所有作者同意投稿，并在相关文件中署名。
- 在线投稿：将论文和相关附件上传至期刊的在线投稿系统，按照系统提示完成投稿过程。在提交之前，检查所有文件是否已正确上传，确保您提交的版本是最终版。
- 等待审稿结果：投稿后，您需要等待期刊的审稿过程。审稿周期因期刊而异，可能需要几周到几个月的时间。在此期间，保持对投稿系统的关注，随时了解审稿进度。
- 回复审稿意见：如果收到审稿意见，请仔细阅读审稿人的建议，并按照要求进行修改和回复。保持礼貌和专业，对审稿人的意见表示感谢。修改后，将修订后的论文和回复函上传至投稿系统。
- 等待最终决定：在您回复审稿意见并提交修订稿后，期刊编辑会根据审稿人的意见和您的回复做出最终决定。决定可能包括接受、接受前需进行进一步修改或拒绝。若论文被接受，按照期刊的要求完成剩余的出版流程。

在整个投稿过程中，请保持耐心和专业态度，根据期刊要求和审稿意见不断改进论文。这将有助于提高论文的质量和被接受的可能性。

2. 投稿注意事项

在投稿科技论文时，注意以下几个方面可以帮助提高论文质量和被接受的机会。

- 选择合适的期刊：在投稿前，务必选择与本人研究领域相关的期刊。关注期刊的影响因子、读者群、出版速度等因素。确保论文满足期刊的格式和内容要求。
- 遵循学术道德：确保本人的研究遵循学术道德规范，避免抄袭、篡改数据或其他不道德行为。对他人的研究成果给予适当引用和致谢。
- 语言表达清晰：确保论文的语言清晰、简洁、规范，避免使用冗长、复杂的句子。对于非英语母语作者，可以寻求英语母语人士的帮助，以确保论文的语言质量。
- 详细撰写方法和结果：方法和结果部分应详细描述实验设计、数据收集和分析方法，以及实验结果。这有助于读者理解并重现本项研究。
- 图表和插图质量：遵循期刊的图表和插图格式要求。使用高质量的图表、插图和图片来展示研究结果，确保它们清晰可读，注明来源。

- 参考文献格式正确：按照期刊的要求正确引用参考文献，并确保参考文献完整、准确。使用参考文献管理软件（如 EndNote、Mendeley 等）可以帮助简化此过程。
- 耐心等待审稿结果：审稿过程可能需要较长时间，保持耐心。在收到审稿意见后，认真阅读并根据要求进行修改，保持与审稿人和编辑的良好沟通，积极回应审稿人的意见。
- 认真校对论文：在提交前要认真校对论文的内容和格式，避免低级错误和不符合期刊要求。
- 应对拒稿：拒稿是学术发表过程中常见的现象。如果论文被拒绝，请认真阅读编辑和审稿人的意见，从中汲取教训，改进论文后重新投稿。

总之，在投稿时要认真准备、严谨对待，保持良好的学术诚信和道德操守，才能增加论文被接受的概率，提高个人的学术水平和声誉。

Exercises

Ⅰ. Point out true or false.

1. When submitting to different journals, the format and word count requirements for the paper are the same.

2. Academic search engines can provide journal paper retrieval and viewing functions, but cannot jump to the journal's official website to obtain submission guidelines.

3. Peer experts and supervisors do not have the experience and knowledge of submitting and publishing papers, so they are not suitable for consulting them about submission-related issues.

4. When writing scientific papers, it is necessary to use more vivid and imaginative vocabulary and language as much as possible to attract readers' attention.

5. When submitting to a journal, it is necessary to familiarize oneself with the journal's peer review process and requirements, and adjust the paper's format, word count, citation, etc. according to the requirements.

6. Inductively coupled plasma mass spectrometry (ICP-MS) can detect elements only at high concentrations.

7. Fourier transform infrared spectroscopy (FTIR) is used to measure the absorption of visible light by environmental samples.

8. Nuclear magnetic resonance (NMR) spectroscopy provides information about the molecular structure of compounds.

9. Remote sensing techniques cannot be used to monitor land use changes.

10. Liquid chromatography-mass spectrometry (LC-MS) is not suitable for analyzing polar compounds.

Ⅱ. Answer the following questions according to the text.

1. What are some effective strategies for selecting an appropriate academic journal to submit a scientific paper to ?

2. How can one effectively respond to peer review comments and feedback on a scientific paper ?

3. What are the steps involved in submitting a technology research paper, along with any relevant precautions ?

4. Could you outline the procedure for submitting a scientific paper in the field of technology and provide some key points to be mindful of ?

5. What are some essential factors to take into account when preparing and submitting a technology-focused research paper ?

Ⅲ. Scan the QR code to read the article "Finding New Ways to Get the Measure of Pollution" and then discuss how to select the most suitable equipment for environmental sample analysis to ensure that the need for scientific accuracy is met, while also reflecting the unity between technology and environmental sustainability?

附录

必备词汇

人与自然

adverse factor	有害因素，不利因素
artefact pollution	人工污染
basic national policy	基本国策
birth control	节育
cancerogenic effect	致癌作用
chronic disease	慢性病
city classification	城市分类
civic-minded	有公德心的
conventional industry	传统产业
conservation of energy	能量守恒
conservation of mass	质量守恒
law of conservation of energy	能量守恒定律
law of conservation of mass	质量守恒定律
coordination theory of development and environment	发展与环境协调论
density of population	人口密度
dust storm	沙尘暴，沙暴
element geochemistry	元素地球化学
Ency	百科全书
Ency Brit	英国百科全书
endemic disease	地方病
environmental accident	环境事故
environmental capacity	环境容量
environmental evidence	环境标志
archives of environmental protection	环保标志

product quality mark	产品质量标志
environmentalist	环境学家，环境工作者，环境专家
environmentally sound	对环境无害的，合乎环境要求的
environmental self-purification	环境自净
environmental technology profile (ETP)	环境技术概貌
epidemic	流行病
geosphere	陆界，陆圈，岩石圈，地（球）圈
social sphere	社会圈
hygiene standard	卫生标准
infant mortality	婴儿死亡率
instinct	本性，本能
Kyoto protocol	京都议定书（《联合国气候变化框架公约的京都议定书》）
lead poisoning	铅中毒
malignant	恶性的
national park	国家公园
population migration	人口迁移
primary pollutant	一次污染物，原生污染物
public health	公共卫生
residential areas	居民区
residential district	居住区
short sight	近视
solar power	太阳能
theory of treatment after pollution	先污染后治理说
United Nations Environment Programme	联合国环境规划署
United States Environment Protection Agency	美国环境保护局
urban community	城市社区
urban disease	城市病
urban heat island	城市热岛
World Food Day	世界粮食日

生态环境与修复

alkaloid	生物碱
amphibious	两栖的
terrestrial animal	陆栖动物

agrobiology	农业生物学
enzyme	酶，化学酶
antioxidant enzyme	抗氧化剂酶
marker enzyme	标志酶
antipollution	防污染，去污染
artifical vegetation	人工植被
background value of soil	土壤背景值
edaphic	土壤的，土壤层的
long (itudinal) profile	纵剖面
elevation profile	纵剖面
pedosphere	土壤圈，土界
reclaim	（土地的）开垦，改良，回收
soil bacteria	土壤细菌
soil conditioner	土壤调理剂
soil fertility	土壤肥力
soil horizon	土层
bacterium	细菌
base manure	基肥
coated granular fertilizer	包膜肥料
compound fertilizer	复合肥料
farmyard manure	农家肥料
mineral fertilizer	无机肥料
natural fertilizer	天然肥料
bioaccumulation	生物累积
biological pest control	生物防治
blight	枯萎病
carbonnitrogen cycle	碳氮循环
cell wall	细胞壁
chlorophyll	叶绿素
community ecological	生态群落
antagonistic function	拮抗作用
competitive effect	竞争作用
synergism	协同作用
complex sample	复合试样
conservation area	保护区

conservation of natural resource	自然资源保护
nature conservation	自然保护
habitat	生境，生境会议
ecological consciousness	生态意识
ecological disruption	生态失调，生态破坏
ecologic disturbance	生态失调
habitat conservation	生境保护
habitat destruction	生境破坏
ecologic(al)	生态的，生态学的
niche	小生境，适当的位置
cropping system	耕作制度
fallow land	休耕地
deforestation	砍伐森林，清除森林
forest deterioration	森林破坏，森林荒废
man-made forest	人工林
temperate rainforest	温带雨林
virgin forest	原始森林，未来发的森林
degree of stability	稳定度
destructive effect	破坏作用
encroach	侵入，侵占，侵蚀
enrichment culture	富集培养
environgeology	环境地质学
exposure test	暴露试验
field (moisture) capacity	田间持水量，田间保水量
field survey	现场调查，实地调查
food crisis	粮食危机
germination	发芽
green area	绿化区
growth curvature	生长曲线
orticulture	园艺学
index species	指标生物
insect biology	昆虫生物学
ex situ conservation	迁地保护
in situ	就地
in situ bioremediation	原位生物修复

integrated control of biological pollution	生物污染综合防治
intercellular	细胞间的
land conservation policy	土地保护政策
lignin	木质素，木素
marginal effect	边缘效应，边界效应
microstructure	微观结构，显微结构
mildew proof	防霉，不生霉
mildew stain	霉斑
nutrition and health care	营养与保健
organic agriculture	有机农业
organic contaminant	有机污染物
organic pesticide(s)	有机农药
passivating agent	钝化剂
percentage of moisture	含水率
percentage of saturation	饱和率
photosynthetic	光合的
plant ecology	植物生态学
pollution free vegetables	无公害蔬菜
root environment(of plants)	植物的根系环境
salinity	盐度，含盐量
salinity control	盐度控制
selective enrichment	选择性富集
sieve	筛
sieve mesh	筛网
submersed plant	沉水植物
symbiotic	共生的
synecology	群落生态学，群体生态学
translocation of plant	植物的转移
transpiration	蒸腾作用，流逸
trophic level	营养级，营养水平
unconfined aquifer	自由含水层
vicious cycle of agroecosystem	农业生态恶性循环
water and soil conservation	水土保持
water content	含水量

weed control	杂草控制
wetland	湿地
wildlife conservation	野生生物保护
winter resistance	抗寒性，耐寒性

大气与大气污染

air mass	气团
air plankton	空气浮游生物
arterial traffic	干线交通
atmosphere	大气，大气压，大气圈
atmosphere pollution	大气污染
atmospheric environmental capacity	大气环境容量
atmospheric pollution indicator plant	大气污染指示植物
atmospheric quality standards	大气质量标准
ozonosphere	臭氧层
chemosphere	臭氧层，光化（大气）层
knoisphere	尘圈，尘层
normal atmosphere	标准大气压
troposphere	对流层（0～10km）
stratosphere	平流层同温层（10～50km）
mesosphere	中和层（50～90km）
thermosphere	热层（电高层）(100km 以上）
bag filter	袋式过滤器
biological gas	沼气
city fog	城市雾
dense fog	浓雾
dust fog	尘雾
city gas	城市煤气
cloth envelop collector	布袋过滤器
fabric filter	布袋过滤器
continuous diffusion	连续扩散
cyclone	旋风（分离，除尘）器
tubular cyclone	管状旋风除尘器
electrostatic precipitator	静电除尘

scrubber	洗涤器
damp air	湿空气
entrain	夹带，输送，吸入
entrained air	夹带的空气
moist air	湿空气
pure air	纯空气
scavenging air	清净空气
secondary air	二次空气
fire damp	沼气，甲烷
forced convection	强制对流
heat convection	热对流，对流换热
heat island effect	热岛效应
heatproof	隔热的，耐热的
incidental release	意外释放，意外排放
integrated control of atmospheric pollution	大气污染综合防治
light pollution	光污染
moderator	缓和剂，减速剂
ozone	臭氧
rainfall	雨量，降雨
rain(-)out	云中的尘埃被形成云的水滴清除
run(-)off	径流，径流量，流失
wash out	云下洗脱
scrubber wash tower	洗涤塔
soot	烟粒
suction	吸气，吸入
suction inlet	吸入口
tail gas treatment	处理
thermal island	热岛（指市中心区气温偏高的地区）
ventilating chamber	通风室
volatile organic compounds (VOC)	挥发性有机化合物

水体与水体污染

primary treatment	一级处理
screen(ing)	筛分
fine screen(ing)	精筛，细筛

screening test	预备实验
classifying screen	分级筛
sedimentation	沉淀
sedimentation basin	沉淀池
flotation	悬浮
flotation tank	浮选池，浮选槽
oil separation	油液分离
equalization	均等，调匀，补偿，稳定
neutralization	中和
grease removal tank	除油池
reservoir	储存库，水库，吸收库，储层，蓄水库
retention	截留，保留
retention basin	水池，储留池
secondary treatment	二级处理
aerobic treatment	好氧处理
anaerobic treatment	厌氧处理
activated sludge process	活性污泥处理
sequencing batch-flow reactor activated sludge process (SBR)	间歇式活性污泥法
complete mix activated sludge process	完全混合活性污泥法
biological filter	生物滤池
biological fluidized bed	生物流化床
extended aeration (or total oxidation) process	延时曝气法
contact stabilization	接触稳定法
wastewater stabilization ponds	废水稳定池
trickling filters	滴滤池；生物滤池
central wastewater treatment	污水集中处理
clarifying basin	澄清池
collecting basin	（集）水池
facultative aerated lagoon	兼性曝气氧化塘
lagoon	氧化池，环礁湖
aerated lagoon	曝气塘
stabilization pond	稳定塘
oxidation channel	氧化沟
oxygen deficit	氧亏

rapid filter	快滤池
slow filter	慢滤池
tertiary treatment (advanced treatment)	三级处理
activated carbon adsorption	活性炭吸附
ion exchange	离子交换
reverse osmosis	反渗透
electodialysis	电渗析
baffle	挡板，折流板
baffle washer	折流洗涤器
cross baffle	折流板
safety screen	安全挡板
biochemical oxygen demand (BOD)	生物需氧量
carrying capacity of water environment	水环境容量
chemical oxygen demand (COD)	化学需氧量
dissolved oxygen (DO)	溶解氧
dissolved solid (DS)	溶解固体
total organic carbon (TOC)	有机碳总量，总有机碳
total oxygen demand (TOD)	总需氧量
total suspended participate (TSP)	总悬浮微粒
total suspended solid (TSS)	总悬浮固体
average sewage	一般污水，中等程度污水
crude sewage	原污水
raw sewage	原污水，未经任何处理的污水
raw sludge	原污泥
digested sludge	消化污泥
domestic sewage	生活污水
flocculation sludge	絮凝性污泥
industrial sewage	工业污水
irrigation with sewage	污水灌溉系统
settled sewage	澄清的污水
settled sludge	沉积的污泥
settling pond	澄清池
sewage pond	污水塘
sewage pump	污水泵
sludge thicking	污泥浓缩

batch treatment	批处理
aquatic environment	水生环境
available oxygen	有效氧
boiler system	供热系统
buffer action	缓冲作用
chilling	冷却，冷凝，淬火
domestic water system	生活给水系统
egress and ingress	出入
emission level	排放标准，排放水平
equilibrium time	平衡时间
eutrophication	富营养化，沃化，加富过程
feed water treatment	给水处理
fermentation alcohol	发酵酒精
filter press plate	压滤机板
floatation pond	气浮池
flocculant	絮凝剂
fresh water	新鲜水，淡水
fresh water pollution	淡水污染
ground water	地下水，地下水位，潜水
hand cleaner	洗手剂
hard water	硬水
hollow fiber	中空纤维
impoundment	蓄水，积水，人工湖，蓄水池
integrated control of water pollution	水污染综合防治
intermittent filter	间歇式过滤机
jacket	夹套
jacketed heat exchanger	夹套式换热器
joint cross	十字接头
joint ring	垫圈
linear adsorption	线性吸附
linear source of water pollution	水污染线源
liquid relief	安全泄液
loss of head	压头损失
mixed bed	混合床
municipal waste distribution systems	城市配水系统

ocean pollution	海洋污染
output	输出，产量，排出量
packed bed filter	填充层过滤器
packing house waste water	食品加工废水
pressure tank	压力水箱
red tide	赤潮，红潮
reuse water	回用水，再生水
running piping	输送管道
sanitary standards for drinking water	生活饮用水卫生标准
scale deposit	积垢，水垢
scale effect	尺度效应，放大效益
surface runoff	地表径流
system of water supply	给水系统
turbidimeter	浊度计
vessel waste	船舶废弃物，船舶废水
void water	孔隙水
wastewater irrigation	污水灌溉
waterproof	不透水的，防水的

声与噪声污染

attenuation factor	衰减因子
audibility threshold	听阈
audio noise meter	噪声计
fatigue auditory	听觉疲劳
ground noise	本底噪声
hearing loss	听力损失
industrial noise	工业噪声
integrated control of noise pollution	噪声污染综合防治
noise background	噪声本底
regenerated noise	再生噪声
resonance effect	共振效应
sound barrier	声屏障
sound eliminator	消声器

sound intensity	声强
sound level	声级
vibration criteria	振动评价标准
vibration proof	防振的，耐振的

固体与固废处置

ash	灰分
biodegradation of refuse	垃圾生物降解
bog muck	泥炭
boiler dust	锅炉粉尘
charcoal	木炭，活性炭
combustibility	可燃性
flammability	可燃性
ignition	点火，点燃
ignitability (I)	可燃性
corrosivity (C)	腐蚀性
reactivity (R)	反应性
toxicity (T)	毒性
inflammability	可燃性，易燃性
pilot	飞行员，领航员，引水员
pilot flame	引燃火焰
danger threshold	危险阈，安全限值
deflagrability	爆燃性
dioxin	二噁英
domestic garbage	生活垃圾
fecal pollution	粪便污染
fire chamber	燃烧室
firming agent	固化剂
food waste	食余残渣，食物废物，厨房垃圾
furnace slag	炉渣
garbage disposal	垃圾处理
high-energy fuel	高能燃料
hospital sewage treatment	医院污水处理
hospital waste	医院废物

household waste	家庭垃圾，生活垃圾
jaw crusher	颚式破碎机
landfill of refuse	垃圾填埋
litter bin	垃圾箱
low temperature separation	低温分离
luminance	发光度
luminous environment	光环境
mine refuse	矿渣
mine waste	矿山废物
municipal waste	城市废物
nightsoil sludge	人粪尿污泥
ocean disposal	远洋废物处理
personal protection	个人防护
radiation loss	辐射损失
radioactive isotope	放射性同位素
refuse disposal	垃圾处置
sanitary landfill	卫生填埋
slag	炉渣
slagging	造渣
solid waste disposal	固体废物处置
tailings	尾渣，尾矿，尾馏分，谱尾
tank waste	废料场
vertical	垂直的，垂直线，铅直的
vertical chamber oven	立式炉
waste minimization	废物最小量化
waste recover	废物回收
waste reduction at the source	废物源头削减

样品前处理与分析

air oven	烘箱
baking oven	烘箱
air sampler	空气取样器
continuous sampling	连续采样
environmental sampler	环境取样器
sample cell	吸收杯，吸收样品池

sample deviation	样本偏差
sample point	取样点
sampler	取样器，进样器
culture dish	培养皿，彼德里氏皿
disinfection plant	消毒设备
laboratory	实验室
labware	实验室器皿
ultrafilter	超滤机
piston	活塞
piston pump	活塞泵
valve end	阀头
vacuum chamber	真空室，真空箱
vacuum pump	真空泵
vacuum seasoning	真空干燥
wash bottle	洗瓶
atomic fluorescence spectrometry	原子荧光光谱法
anode stripping voltammetry	ASV 阳极溶出伏安法
auxiliary apparatus	辅助设备
capillary column	毛细管柱
chromatographic adsorption	色谱吸附
gas chromatography	气相色谱法，气相层析
coextraction	共同萃取
electroanalytical chemistry	电分析化学
express-analysis	快速分析
extraction	萃取，提取
reextract(ion)	反萃取
flame atomic absorption spectorphotometer	火焰原子吸收光谱测定法
fractional device	分馏装置
high performance liquid chromatography(HPLC)	高效液相色谱法
inductively coupled plasma	电感耦合等离子体
infrared	红外线的
instrumental analysis	仪器分析
ion exchange chromatography (IEC)	离子交换色谱法
mass spectra	质谱
nephelometric analysis	浊度分析

photoelectric analysis	光电分析
spectral absorption	光谱吸收
spectrophotometer	分光光度计
temperature programmed chromatography	程序升温色谱法
total ionic strength adjustment buffer	总离子强度缓冲剂
ultramicroanalysis	超微量分析
ultrasonic extraction	超声萃取
visible spectrometry	可见光分光法
X-ray absorption analysis	X-射线吸收分析法
refraction	折射
supercritical fluid extraction	超临界流体萃取
analytical pure	分析纯，二级纯
chemically pure reagent(CP)	化学纯（三级品）试剂
industrial grade	工业级
macroanalysis	常量分析
trace level	痕量级，示踪量
ultramicro	超微量，超微
ultra trace	超痕量
ultratrace analysis	超痕量分析
small-scale test	小型试验
bench-scale	小型的，实验室规模
bench test	小型试验（比实验室试验大，比中型试验小）
air seal	气封
water seal	水封
application standard	应用标准
approval test	鉴定实验
aseptic culture	无菌培养
baseline	基线
between-cluster variance	组间变异
between-strate variance	区间变异
check sample	对照试样，核对试样
check test	对照试验，核对试验
chi-square distribution	卡方分布
coefficient of correlation	相关系数
coefficient of diffusion	扩散系数

coefficient of recovery	回收系数
competitive effect	竞争作用
control	控制，管理，防治
control analysis	控制分析
control experiment	对照试验，检验试验，控制试验
control sample	对照试样
correction of error	误差校正
correlation analysis	相关分析
critical coefficient	临界系数
dead time	停机时间，停歇时间
detection limit	检出限
deviation	偏差，误差，偏斜，偏离，偏差数
edge zone	边缘地带
effective age	有效使用期
elephant	大象，绘图纸，起伏干扰
eluent gas	洗脱（用）气体，载气
end effect	末端效应
evaluation criterion	评价标准
experimental model	试验模型，实验模型
free settling	自由沉降，自然沉降
gene expression	基因表达
gravity	重力，地心引力
gravity settling	重力沉降
hypothesis test (ing)	假设检验
interim trial	临时试验
internal standard	内标，内标物，内标准
ligand	配体，培基
ligand exchange	配体交换
limit	极限，范围
limit analysis	极限分析
linked reaction	偶联反应
liquor supernatant	上层清液
major constituent	主成分
major factor	主要因素
masking	掩蔽，隐蔽

masking agent	掩蔽剂
masking effect	掩蔽效应
matrix interferece	基体干扰
maximum error	最大误差
maximum temperature	最高温度
maximum thermometer	最高温度计
mean deviation	平均偏差，平均偏移
mean value	均值
measuring accuracy	计量精度
median lethal concentration (LC50)	半数致死浓度
median lethal dose (LD50)	半数致死剂量
method of standard addition	标准物添加法
methodology	方法论，方法学
monitored control system	检查控制系统
monitoring equipment	监控装置
numerical analysis	数值分析
observation method	观察法
observation error	观察误差
one factor variance analysis	单因素方差分析
partial saturation	部分饱和
peak-to-peak value	峰值到峰值，p-p 值，峰-峰值
qualitative test	定性试验，定性测定
quality arbitration	质量检定
quality assessment	质量评价
quantitative reaction	定量反应
reaction tower	反应塔
reactive resin	活性树脂，反应型树脂
receiver	接收机，收集器，储液罐，转化炉
rectification column	精馏塔
regression	回归
regression analysis	回归分析
relative deviation	相对偏差
relative toxicity	相对毒性
removal efficiency	去除系数，去除效率
residual analysis	残差分析

routine check	日常检验
run in test	空转试验
significance test	显著性检验
slime index	黏泥指数
split ratio	分流比
standard addition method	标准加入法
standard methods of analysis	标准分析法
t-distribution	*t* 分布
T-test	*T* 检验
valid figure	有效数字
variation	方差
visibility	可见性，可见度
void volume	空隙容积，空隙率
volumetric error	滴定误差
weighted factor	权重因数

化学基础

anesthetic action	麻醉作用
anhydrous alcohol	无水酒精
spirit of wine	酒精
anode	阳极
cathode	阴极
association	缔合作用
carbonification	碳化作用
catchment area	集水面积
cation	阳离子，正离子
cell	细胞，比色皿，电解，电池
cell differentiation	细胞分化
cell-driven vehicles	电动汽车
cell length	比色皿长度
cell line	电解槽系列，细胞系
cell room	电解车间
charge balance	电荷平衡
chelant	螯合剂
chelating agent	螯合剂

chemical passivity	化学钝性
chemical process	化工工艺，化学加工
conductance water	电导水
endocrine	内分泌
functional group	官能（原子）团
general chemistry	普通化学
graphite	石墨
halogen	卤素，卤
heat treatment	热处理
heavy metal	重金属
heavy metal pollution	重金属污染
histology	组织学
impurity	杂质
background impurity	本底杂质
extrinsic contaminants	外来杂质
foreign impurity	外来杂质，夹杂物，异物
foreign matter	杂质
tramp material	外来杂质
microimpurity	微量杂质
latitude	纬度，宽容度
latitudinal distribution	纬度分布
licence	执照，许可证，特许
licence system of using water	取水许可制度
macroporous polymer	大孔聚合物
molecular weight	分子量
nitrogen balance	氮平衡
nitrohydrochloric acid	王水
off-grade	等外品
oxidation-reduction reaction	氧化还原反应
passive pollutant	潜在性污染物
perfect gas	理想气体
periodic table	周期表
photoaging	光老化
photolysis	光解（作用）
physical constant	物理常数

physicochemical treatment	物理化学法处理
physiological action	生理作用
part per thousand	千分之……
part per million (ppm)	百万分之……
part per billion (ppb)	亿分之……
part per trillion (ppt)	万亿分之……$(1/10^{12})$
radiation chemistry	辐射化学
radical	根本的，基本的
radical transfer	基因转移
radioactive constant	放射性常数
raw coal	原煤
refrigerant	制冷剂，冷冻剂
refrigerating fluid	冷冻液
regional analysis	区域分析
relative acidity	相对酸度
ideal solution	理想溶液，理想混合物
simple solution	真溶液，分子溶液
solubility	溶解度
specific conductivity	电导率
spirit of salt	盐酸
thermal power station	热电站
thermoanalysis	热分析
thermolysis	热分解
toxic smog	有毒烟雾，毒物
toxic symptom	中毒症状
toxin	毒素
ultrapure water	超纯水
orkability agent	增塑剂，塑化剂
mono-	一
di-(bi-)	二
tri-	三
tetr(a)-	四
pent(a)-	五
hex(a)-	六
octa-	七

enon(a)-	八
hept(a)-	九
deca-	十
meth-	甲基
eth-	乙基
prop-	丙基
but-	丁基
pent-	戊基
hex-	己基
hept-	庚基
oct-	辛基
non-	壬基
dec-	葵基
-ane	烷
-ene	烯
-yne	炔

工业与清洁生产

audit team	审计小组
blower	鼓风机
central station	中心电站
Clean Development Mechanism (CDM)	清洁发展机制
Cleaner Production Promotion Law	清洁生产促进法
discharge fee	排污费
dissertation	学位论文，专题，学术演讲，研究报告
domestic need	国内需求，国内需要
enquiry data	调查数据
environmental audit	环境审计
environmental impact assessment(EIA)	环境影响评价，环境影响评估
Environmental Protection Administration(EPA)	环境保护局
Environmental Protection Agency(EPA)	环保机构
environmental quality	环境质量
environmental quality standard (EQS)	环境质量标准
expected life	预期寿命

hoistway door	电梯厅门
introduction	引言，介绍，倡导，引导
investment project	投资项目
land use pattern	土地利用规划
material balance	物料衡算，物料平衡
middle/high cost CP option	中/高费清洁生产方案
operations research	运筹学
paint coating	涂漆
papermaking	造纸，造纸业
process	过程，方法，工艺
protected landscape	受保护的自然景观
pulp and paper mill wastewater	造纸工业废水
review on special information	专业文献综述
risk evaluation	风险评价
self-acting feed	自动进料
strategical environmental assessment	战略环境评价
system of shutting down and moving polluting enterprises	关、停、并、转、迁移制度
system of “the three at the same time”	“三同时”制度
system of treating environmental pollution within a prescribed time	限期治理制度
zero-pollution goal	零污染目标

参考文献

1. 郁仲莉，王耀庭. 英语写作与翻译实用教程[M]. 2版. 北京：中国农业出版社, 2001.
2. 石立华，苏航. 科技写作方法[M]. 北京：国防工业出版社, 2006.
3. 于苏俊. 环境科学专业英语[M]. 成都：西南交通大学出版社, 2005.
4. 王旭梅，王晓东. 环境科学与工程专业英语[M]. 哈尔滨：哈尔滨工程大学出版社, 2006.
5. 马志毅，苏玉民. 环境保护环境工程专业英语[M]. 北京：中国环境科学出版社, 2003.
6. 蒋东云，李学军. 环境工程专业英语[M]. 武汉：华中科技大学出版社, 2008.
7. 宋志伟. 环境专业英语教程[M]. 哈尔滨：哈尔滨工程大学出版社, 2005.
8. 贺小凤. 室内环境检测专业英语[M]. 北京：化学工业出版社, 2007.
9. 杨维. 给水排水工程与环境工程专业英语[M]. 北京：机械工业出版社, 2009.
10. 钟理. 环境工程专业英语[M]. 北京：化学工业出版社, 2012.
11. 蓝梅. 给排水科学与工程专业英语[M]. 北京：化学工业出版社, 2013.
12. 王春丽. 给排水科学与工程专业英语[M]. 哈尔滨：哈尔滨工程大学出版社, 2016.
13. 孙平. 科技写作与文献检索[M]. 北京：清华大学出版社, 2016.
14. Khoiyangbam R S. Introduction to Environmental Sciences[M]. The Energy and Resources Institute, 2015.
15. Davis M L, Masten S J. Principles of Environmental Engineering and Science[M]. McGraw Hill, 2014.
16. Miller G T, Spoolman S E. Environmental Science[M]. Cengage Learning, 2019.
17. Verma S, Kanwar V S., John S. Environmental Engineering: Fundamentals and Applications[M]. CRC Press, 2022.
18. Miller C T, Gray W G, Bruning K. Evolution of Environmental Engineering: Challenges and Solutions[J]. Journal of Environmental Engineering, 2020, 146(7), 02520001.
19. Rashid G A, Viswanathan K K, Hassan S. The Environmental Kuznets Curve (EKC) and the Environmental Problem of the Day[J]. Renewable & Sustainable Energy Reviews, 2018, 81, 1636-1642.
20. Amaral-Zettler L A, Zettler E R, Mincer T J. Ecology of the Plastisphere[J]. Nature Reviews Microbiology, 2020, 18(3), 139-151.
21. Ouyang Z, Wei W, Chi C G. Environment Management in the Hotel Industry: Does Institutional Environment Matter? [J]. International Journal of Hospitality Management, 2019, 77, 353-364.
22. González-Pinzón R, Dorley J, Regier P, et al. Introducing “The Integrator”: A Novel Technique to Monitor Environmental Flow Systems[J]. Limnology and Oceanography, Methods, 2019,17(7), 415-427.
23. Catford J A, Wilson J R U, Pyek P,et al.Addressing Context Dependence in Ecology[J]. Trends in Ecology & Evolution 2022.
24. Wainwright C E, Staples T L, Charles L S, et al. Links between Community Ecology Theory and Ecological Restoration are on the Rise[J]. Journal of Applied Ecology, 2018, 55(2): 570-581.
25. Mayembe R, Simpson N P, Rumble O, et al. Integrating Climate Change in Environmental Impact Assessment: A Review of Requirements across 19 EIA Regimes[J]. Science of The Total Environment,

2023, 869: 161850.

26. Li J, Pei Y, Zhao S, et al. A Review of Remote Sensing for Environmental Monitoring in China[J]. Remote Sensing, 2020, 12(7): 1130.
27. Hu F, Guo Y. Health Impacts of Air Pollution in China[J]. Frontiers of Environmental Science & Engineering, 2021, 15: 1-18.
28. Lin L, Yang H, Xu X. Effects of Water Pollution on Human Health and Disease Heterogeneity: A Review[J]. Frontiers in Environmental Science, 2022: 975.
29. Ram C, Kumar A, Rani P. Municipal Solid Waste Management: A Review of Waste to Energy (WtE) Approaches[J]. Bioresources, 2021: 16(2), 4275-4320.
30. Mucci N, Traversini V, Lorini C, et al. Urban Noise and Psychological Distress: A Systematic Review[J]. International Journal of Environmental Research and Public Health, 2020, 17(18): 6621.
31. Wu Y, Li X, Yu L, et al. Review of Soil Heavy Metal Pollution in China: Spatial Distribution, Primary Sources, and Remediation Alternatives[J]. Resources, Conservation and Recycling, 2022, 181: 106261.
32. Ong T T X, Blanch E W, Jones O A H. Surface Enhanced Raman Spectroscopy in Environmental Analysis, Monitoring and Assessment[J]. Science of the Total Environment, 2020, 720: 137601.